The Fundamentals of Sound Science

Revised First Edition

By Elena Borovitskaya
Temple University

Bassim Hamadeh, CEO and Publisher
Michael Simpson, Vice President of Acquisitions
Jamie Giganti, Senior Managing Editor
Jess Busch, Senior Graphic Designer
John Remington, Senior Field Acquisitions Editor
Natalie Lakosil, Licensing Manager

First published in the United States of America in 2016 by Cognella, Inc.

Printed in the United States of America

ISBN: 978-1-63487-435-9 (pbk) / 978-1-63487-436-6 (br)

www.cognella.com 800.200.3908

Contents

3. Waves

4. Families of Musical Instruments

8. Perception of Sound

9. Fourier Analysis of Simplest Sound Spectra

10. Basic of Percussion Instruments and Normal Modes

11. Normal Modes of Strings

12. The Violin

13. Flutes and Recorders

14. The Reed Family

Chapter 1

What is Sound?

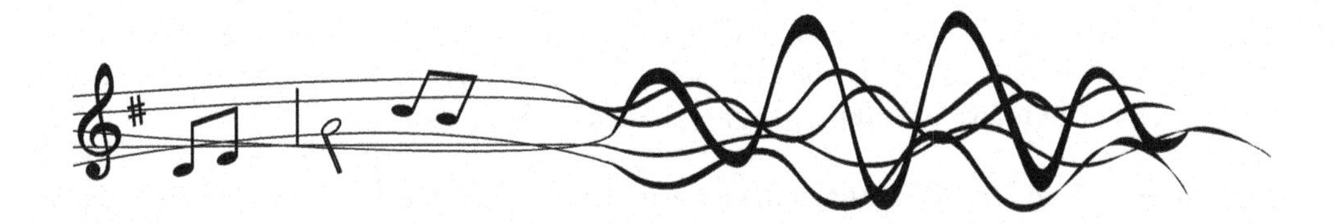

1.1 The Very First Steps

Almost all books about Science of Sound start with the same question: does a tree falling in a forest far, far away produce any sound? Why not begin this book with this age-old question too?

To answer the old riddle, we should first try to give a definition of "sound." And this is a point where the problems immediately begin. We are using the word "sound" to describe at least two different things:

- **Sensation of the auditory nerve of our ear,** means what we actually perceive. From this point of view, the answer to the question would surely be "no." We do not hear a tree falling in the remote forest.

- **The wave** traveling through the medium (for instance, through the air, but not only), which somehow causes the sensation of hearing. The falling tree produces a wave, so from this point of view we should answer "yes" to the age-old question.

All aspects of sound production, traveling, and perception are studied by the science of **Acoustics**.

So, **Acoustics is a science of sound.** Don't be fooled by the simplicity of this definition because Acoustics is an interdisciplinary science, which includes Physics, Engineering, Theory of Music, Psychology, Physiology, Architecture … and maybe something else we have forgotten here: the study of sound that we cannot hear: Ultrasound and Infrasound Acoustics.

1.2 Sound Around Us

We will concentrate our attention on only the audible range of sound for humans. And again we need some definitions. We intuitively distinguish three types of sound around us:

- **Noise** is an unorganized, usually unpleasant sound, often of a very disturbing nature for our hearing.
- **Music** is a combination of sounds that we choose to hear for our enjoyment. Usually music is based on well-defined pattern of sounds. Music, in a practical sense, does not carry information, only emotions; the message of music is very abstract and strongly depends on a lot of an individual properties: mood, background, character.
- **Speech** is also organized, but its main purpose is the carrying of information. Sounds in speech are arranged into words used for communication.

We should understand from the very beginning that the boundary between music and noise is not very distinct and clear. Many admirers of classical opera would consider modern rock music as outrageous noise, and maybe unbearably loud. But it cannot be denied that some pieces by Led Zeppelin and Jethro Tull are now acclaimed classics.

1.3 Distance and Velocity

First, we should get used to some basic concepts. And the very first one is distance traveled. When we answer the question: "How far do you live from campus?" we actually answer the question about distance to be traveled every day in our commute.

For an object that travels at constant speed, for instance, a car moving along a straight road with the cruise control on, distance and speed are connected through a simple formula:

$$\text{distance} = \text{speed} \times \text{time}.$$

We could easily calculate that if we travel at a constant speed of 40 mph for 3 hours, we will cover 120 miles.

The problem is that traveling at a constant speed is a pretty rare situation on any road: our actual speed at any moment depends on traffic, traffic lights, weather conditions, etc. So, the speed during any, even very short, travel is continually changing. We could prove it by watching the speedometer. But still we could use the concept of an average speed of our travel:

$$\text{speed} = \frac{\text{distance}}{\text{total time of travel}} = \frac{d}{t}$$

Distance could be measured using any units of length: meters, feet, inches, miles. The choice of a particular unit depends on the problem we are considering at the moment. It is not very convenient to measure distance traveled by a snail in miles and distance to the Sun in inches.

Because most of the world, and especially scientists, are using the metric system, we will also use meters, for the most part, in this book. The relationships between Imperial and Metric units, as well as prefixes of Metric system, will be considered in Appendix A.

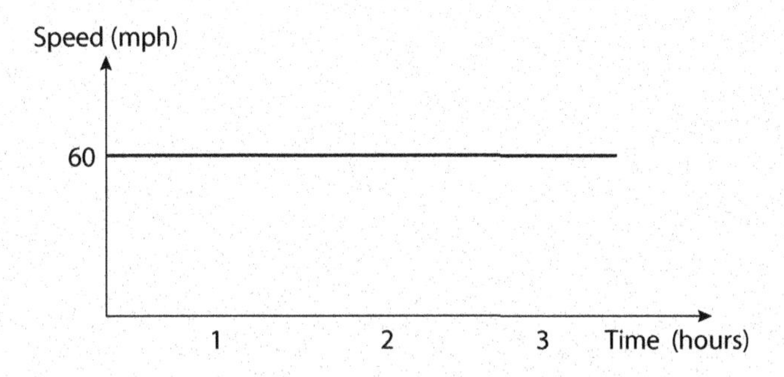

Fig. 1.1. The dependence of speed on time for uniform motion.

Fig. 1.1 shows the dependence of the speed for uniform motion and Fig. 1.2 the dependence of the speed for an object falling with constant acceleration.

Example 1.1. What is the average speed of a car that travels 100 km in 2 hours?

Answer:

average speed = distance/time of travel = 100 km/2 hours = 50 km/h

Example 1.2. A train traveled 100 km in two hours between 9 a.m. and 11 a.m. How fast was this train moving at 10 a.m.?

Answer:

We don't know from the information given. All that we can find is average speed, which is not necessarily equal to the speed at each moment of time.

1.4 Force and Acceleration

Acceleration appears when speed is changing, see, for example, Fig. 1.2. When we start from a traffic light or slow down in traffic, we are changing speed. We could feel it by the reaction of our body in acceleration: while speeding up we are pulled back, while slowing down we are pulled in forward direction. Such reaction of any material object on the change of speed is called inertia.

Average acceleration is defined as the rate of change of speed. In principle, this involves not only the change of magnitude of speed, but also the change of direction of motion, but for our subject it is not important.

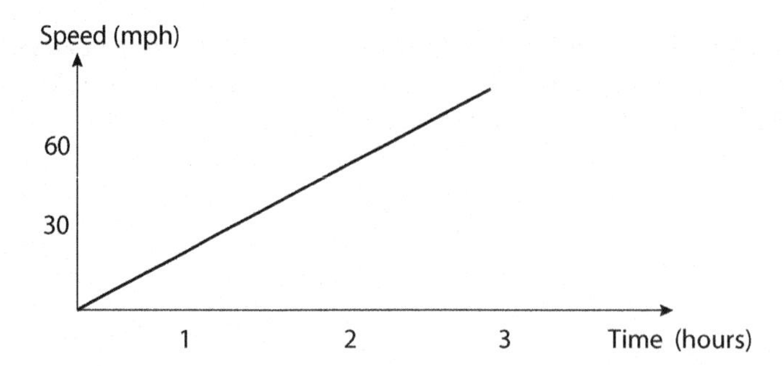

Fig. 1.2. The dependence of speed on time for motion with constant acceleration.

$$\text{acceleration (a)} = \frac{\text{change in speed}}{\text{time elapsed (t)}}$$

So, the keyword for existence of acceleration is **change of speed.**

Example 1.3. A cat starting from rest has speed 4 m/s 2 seconds later. What is the average acceleration of this cat?

Answer:

acceleration = change of speed/time elapsed = (4 m/s – 0 m/s)/2 s= 2 m/s/s (or m/s^2).

Example 1.4. All objects near the surface of Earth are falling with constant acceleration 9.8 m/s^2, which is called the acceleration of free fall. What is the speed of the object released from rest after 2 s of falling?

Answer: This object will gain speed (change speed) at a rate that can be found from:

change of speed = speed final – speed initial = 9.8 m/s^2 x 2s = 19.6 m/s

Because the object starts from rest, the initial speed is 0, so our answer just gives us the final speed after 2 s of free fall.

Concept of force we are using every day, usually describing force as a result of push or pull on an object.

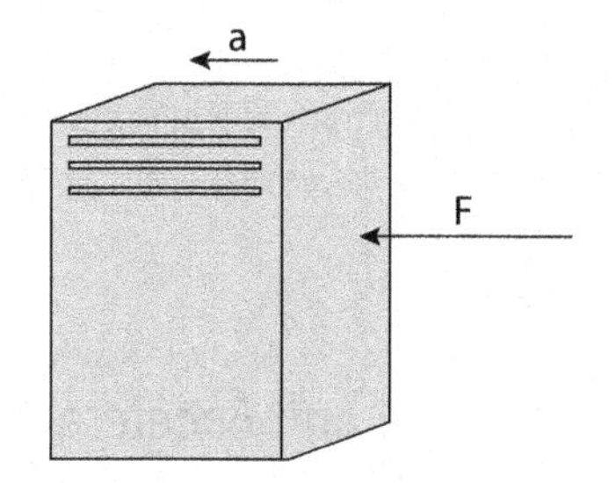

Fig. 1.3. The force exerted on the box results in the appearance of the acceleration in the same direction.

Let us consider a simple example: an object, say a box, (Fig. 1.3) is sitting on a smooth surface. We are exerting force on it, and it accelerates. Intuitively we understand that the bigger effort we apply, the faster will our object accelerate. This can be summarized in Newton's law, which connects force and acceleration:

$$F = m \times a$$

What does it mean? Acceleration is proportional to a force. For instance, twice the force will produce twice the acceleration.

The units of force are newtons (N)

$$1\ \mathrm{N} = 1\ \mathrm{kg} \times 1\ \mathrm{m/s^2}$$

Practically speaking, we always have more than one force being exerted on an object. Let us check Figure 1.4. The object is sitting on a smooth surface at rest. Does it mean that no forces are exerted on it? Definitely not. There is always the force of gravity because of the pull of our planet Earth, and there is also force from the reaction of our surface.

How come those two forces do not accelerate our object? It looks like they cancel each other out to zero. But why?

Any force has two equally important properties:

- **magnitude**, i.e., how much effort somebody (or something) applies to our object; and
- **direction**. The force that acts downward will lead to completely different motion from the force that is acting to the left or to the right.

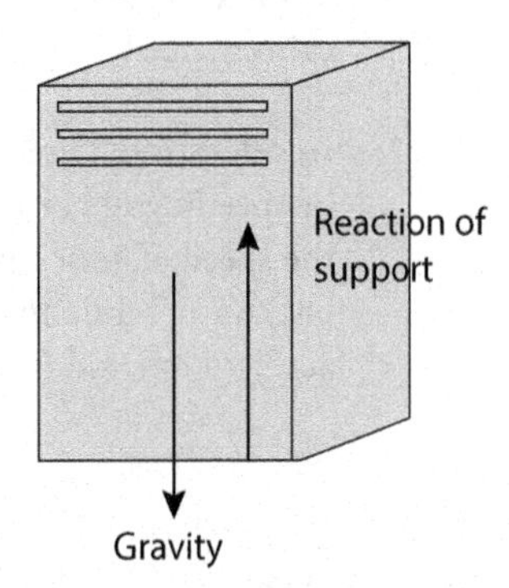

Fig. 1.4. Force of gravity and force of reaction of support cancel are equal in magnitude and of opposite direction. As a result, the box shown is at rest.

In Fig. 1.4 our object is at rest because the force of gravity is exactly equal to the force of reaction of the support in magnitude; these two have opposite directions.

Concepts that have two properties—direction and magnitude—are called vectors. In everyday life we rarely think about so complicated a matter, so this is a reason to consider these concepts in more detail.

1.5 Scalars and Vectors

In this book we will use two flavors. The first one, called scalar, has only one important characteristic: magnitude, which is just a number. Examples of scalars are mass, temperature, volume. There is no direction for a mass of your body; this is just a number that you see while standing on a bathroom scale.

Time is also a scalar, although this is not as obvious as for mass. Time does not have a direction, despite many sci-fi movies that like to discuss "the vector of time," whatever that means.

Vectors are pretty often encountered in Physics. These have two equally important properties: magnitude and direction. They are a little bit more sophisticated for students, yet could be represented pretty easily as an arrow of a length corresponding to the magnitude of a vector.

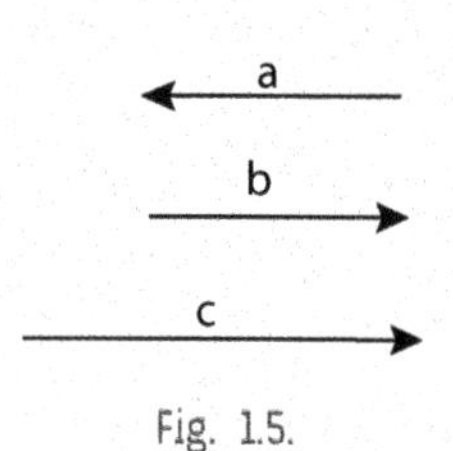

Fig. 1.5.

This arrow points in the direction of our vector. In Fig 1.5 are shown three vectors. Vector (a) and (b) have the same magnitude but different directions. So, we cannot say that (a) and (b) are equal. Vectors (b) and (c) have the same direction but different magnitude, which also makes them not identical.

From the very beginning it is good to understand that direction of a vector is as important as magnitude.

What vectors do we deal with every day?

1. **Velocity**. It has direction, for instance, north or east, and magnitude, which we are just calling speed.

2. **Acceleration**. Direction of acceleration you "feel" while slowing down at a traffic light, or speeding up, or making a turn. Just analyze your sensation when you are changing the speed of your car. While slowing down, your body is pulled forward; speeding up, backward; while making a turn, you are pulled outward. Any object tries to "resist" acceleration and is pulled in the direction **opposite** the existing acceleration.

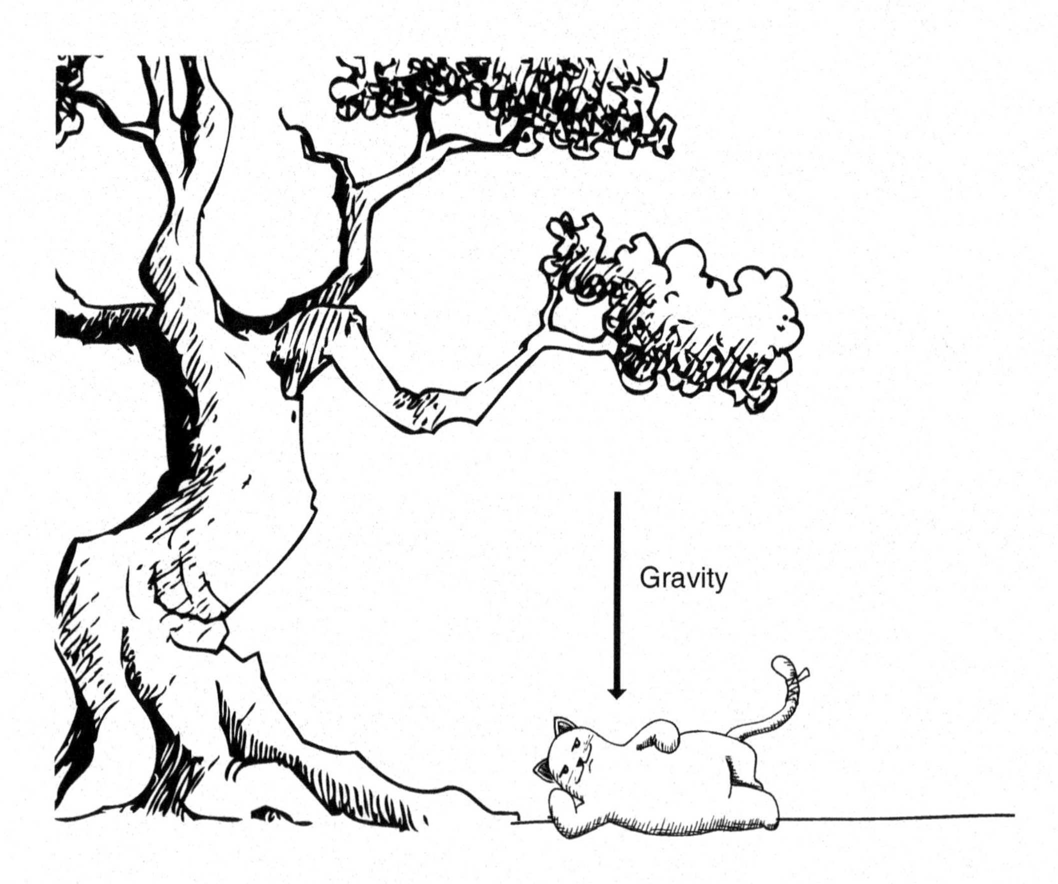

Fig. 1.6a. Force of gravity acting when an object is in contact with ground.

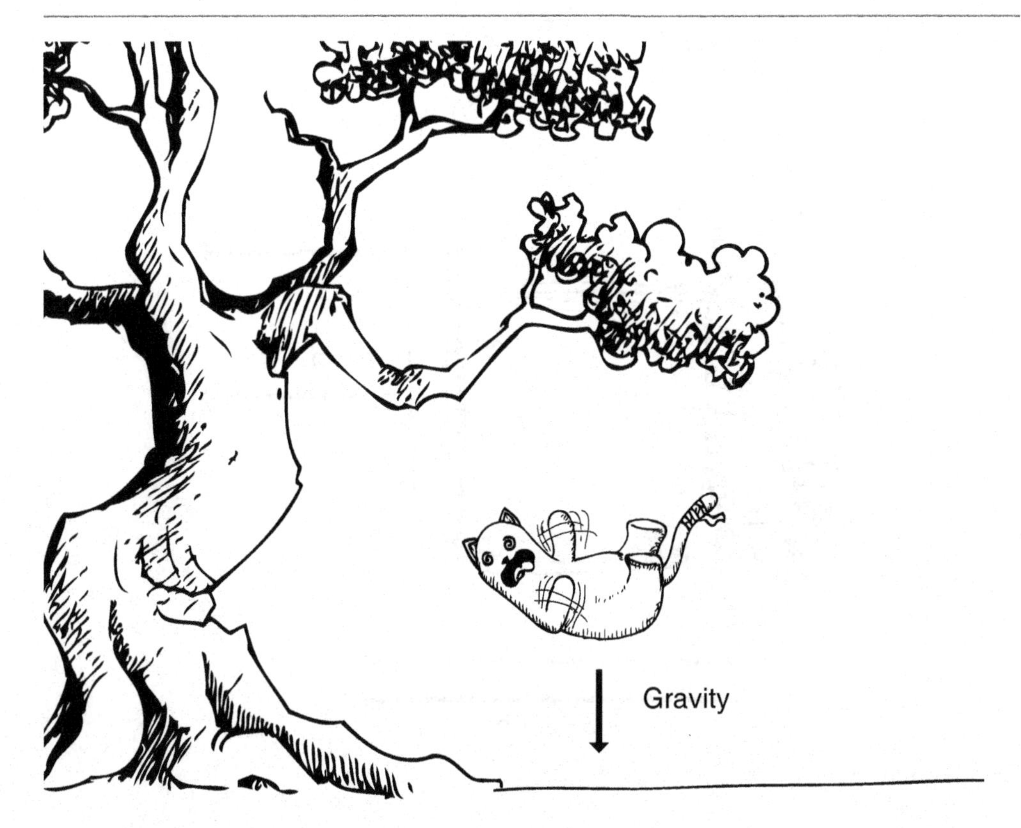

Fig. 1.6b. Force of gravity acting on a distance.

3. **Force.** As we already saw from Fig. 1.4, force of gravity has the opposite direction to a force of reaction of support. As a result, these two—maybe even pretty big forces—cancel each other, and our object is at rest; it does not accelerate. This simple example shows us that while acting with vectors we should be very specific, keeping in mind not only magnitude of force, but also direction.

When we consider a force, we should always understand who (or what) is exerting this force on our object. There are **only a few** forces that can be exerted on a distance without contact. One of them everybody knows very well. It is force of gravity. This force is exerted on us and any object around by the Earth. You could be in contact with our planet, lying on the grassy backyard, or you could be not in contact at all, jumping, for instance, from the branch of a tree, but force of gravity is exerted on you in both of these situations (see Figs. 1.6a and b).

Most forces can be exerted only with contact. Such force disappears at the moment of losing contact. Let us go back to Fig. 1.4. If your object rolls from the edge of this support, the force of reaction will immediately disappear. There will be only one force present, force of gravity, under the influence of which our object will accelerate downward.

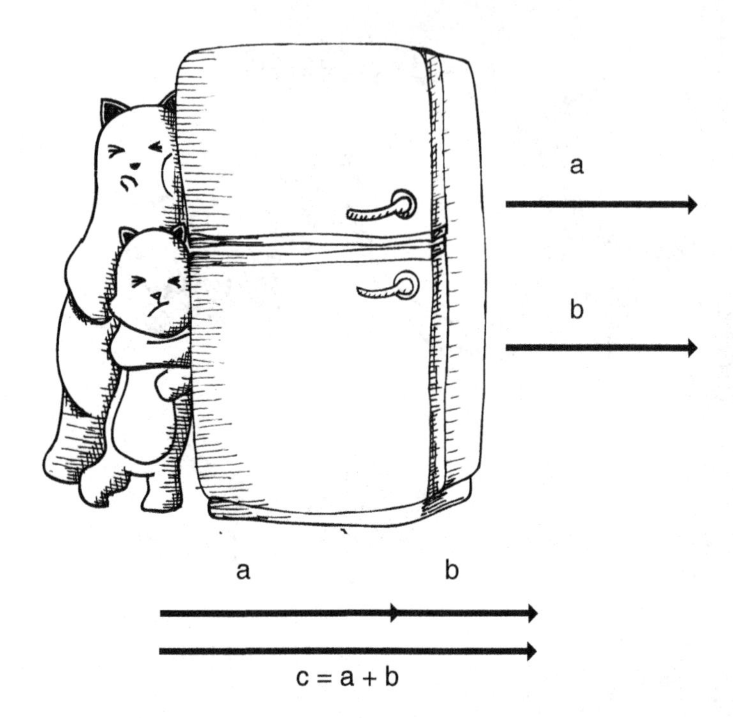

Fig. 1.7a.

Fig. 1.7b.

1.6 Addition of Vectors

We should consider some operations with vectors to better understand what result we could expect while applying two different forces to an object. Let us imagine a situation where we are pushing a refrigerator with our friend.

First we will try to push our object in the same direction as shown in Fig. 1.7a. The resulting force will have the same direction as either of the two (they are parallel) and magnitude, which is just a simple sum of two forces. This situation has no difference when adding two scalars: the mass of two objects is just a sum of the masses. It is important to emphasize that in the situation of two vectors parallel to each other, we are getting the **maximum** possible result.

People are doing strange things, so next time we will push our refrigerator in opposite directions, as shown in Fig. 1.7b. The resulting net force will have the direction of the bigger force of the two acting and a magnitude equal to the difference between them. So, you could exert huge forces, but result will be defined by a difference. If you are applying equal forces, the difference is equal to zero and your object will not move at all. In this situation of antiparallel vectors, we are getting the **minimum** possible result of the addition of two vectors.

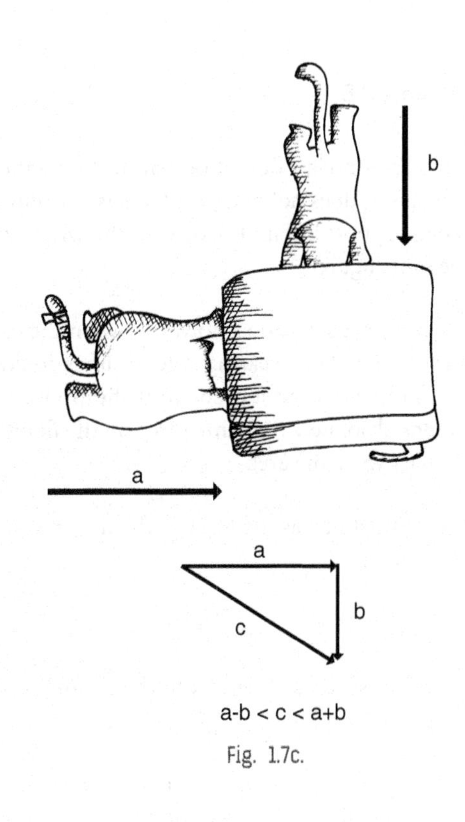

Fig. 1.7c.

Finally, we will exert two forces that are neither parallel nor antiparallel to each other. Say, perpendicular as shown in Fig. 1.7c. The net force in such a situation in **not parallel** to any of our forces, and its magnitude will be somewhere in between its maximum and minimum value. Where in the range it is depends on the angle created by two forces.

This simple example shows us that if we have two vectors and change the direction of only one of them, we could get **completely different results**.

Example 1.5. We have two vectors of magnitude 3 and 4 units.

a. What mutual orientation will give the maximum possible result of the addition of these two vectors?

Answer: The maximum possible result will happen if these two vectors are parallel to each other as shown in Fig. 1.7 a. The length of the resulting vector is 3 + 4 = 7 units.

b. What mutual orientation gives the minimum possible result of the addition of these vectors?

Answer: The minimum result can be obtained if these two vectors are antiparallel to each other as shown in Fig. 1.7 (b). The length of the resulting vector is 4 - 3 = 1 unit.

1.7 Pressure

Newton's second law of motion tells us that if there is a net force acting on an object, this object accelerates. It does not matter if force is distributed over some surface, such as the force of reaction of a support in Fig. 1.4, or over the whole volume of an object, such as the force of gravity in the same figure.

There are some situations, however, when the distribution of forces matters. Everybody knows that even a tiny lady wearing stiletto shoes could leave marks on a hardwood floor; meanwhile, there will be not such a problem if she would wear sneakers. But this lady is exerting in both situations absolutely the same force on the floor. So, there is only one difference: distribution of this force over some area.

Pressure is defined as the force divided by the area over which this force is distributed, or force per unit area:

$$p = \frac{\text{force (F)}}{A}$$

Units of pressure, as we see from the definition, are units of force divided by units of area, 1 Pa (pascal) = 1 N/m^2.

The concept of pressure is especially important when we consider phenomena in fluids (liquids and gases). Fluids exert forces on the walls of their containers and anything immersed in them. The pressure at any point in a container of fluid is determined by the fluid above this point. For example, the Earth's atmosphere above us results in a pressure of about 10^5 N/m^2 at sea level, which is taken as another unit of pressure, 1 atmosphere (1 atm), and is useful when discussing phenomena in gases. If we travel up to the mountains, the pressure exerted on us will gradually decrease. At the altitude of 6.5 km the pressure is only 0.5 atm.

Example 1.7. What force is exerted on your chest by atmospheric pressure? Why does this force not crush your chest?

Answer:

$$F = P \times A = 10^5 \text{ Pa} \times 0.5 \text{ m} \times 0.5 \text{ m} = 25\,000 \text{ N},$$

where we estimated the size of our chest as the square with sides 0.5 m, which is a pretty fair estimation. Note, that the force 25 000 N corresponds to a weight of 2500 kg (5500 lbs). This force does not crush your chest only because the pressure on the outside is equalized by the pressure of bodily fluid on the inside.

The concept of pressure is extremely important for the Science of Sound. Sound waves are waves of pressure, which means they consist of small and rapid variations of pressure in the air. The amplitude of these variations defines loudness of sound, and rapidness—the pitch of sound.

1.8 Work and Energy

The key concepts for understanding any physical phenomena are work and energy. These words we all know from everyday life, but in Physics they have very strict definitions.

Work is done any time a force is applied to an object that moves. The work done by a constant net force is, in the simplest situation, a product of a force times the distance moved parallel to the force:

$$W = F \times d$$

The units of work are newtons-meters, or joules (J). If we want to lift an object of weight 100 N to the height of 2 m, we should do the work 100 N x 2 m = 200 J.

Energy is defined as ability to do **work**. We lifted a weight of 100 N in the above example to the height of 2 m, doing 200 J of work. Energy is measured in the same units as work; as a result, 200 J of energy are now stored in this object. We could release it from this height, and it will do some useful work.

In our study of sound we are concerned mainly with mechanical energy. Sound waves carry mechanical energy. A vibrating system, such as a vibrating string, has mechanical energy. It is doing work on molecules of air, setting them into motion, and they, in turn, are setting into motion their neighbors. Energy passes from one molecule to another, and finally the molecules adjacent to your eardrums set them into vibration. All these objects are doing work on each other, delivering the energy from the source to the receiver—your auditory nerve and finally to the brain.

It is worth noting that energy cannot be magically created or destroyed. It can only be transferred from one object to another, or it could change into other forms, which may not be as useful as mechanical energy. If we release the above object of the weight 100 N from the height of 2 m, it will accelerate, hit the ground, and stop. Does its energy disappear? No. This energy simply changed from the form of mechanical energy of motion to thermal energy: the temperature of the object will be slightly higher than before.

For purposes of our study of sound, let us describe two types of mechanical energy.

1. **Kinetic energy** is energy of motion. Any moving system, such as a car on a road, vibrating string or drum, or molecule in the air, has kinetic energy, which is defined as

$$KE = \frac{mv^2}{2}$$

where m is mass and v the speed of our object. If the object that you consider has mass and speed, it has energy of motion.

2. **Potential energy** is the energy of configuration of objects in our system. The definition of potential energy strongly depends on the forces acting inside of our system.

Potential energy of an object lifted above the ground is defined by the force of gravity. Look again at this example: you did not change your object, you just lifted it; you did not do anything to our planet Earth, which is providing the force of gravity. But you changed the **mutual** configuration of these two, and, as a result, your system got some stored potential energy.

The potential energy of a stretched string is defined by the force of tension. Let us consider a string on a guitar. If we do not touch the strings, they are at rest. Now we will move one of the strings aside, stretching it. The string is still at rest, but as a result of stretching, we have now some potential energy stored in it. Again, all we did was to change the configuration of objects inside of our system.

The potential energy associated with different forces should be defined specifically for any situation. For example, the potential energy of an elastic spring will depend on the stiffness of this spring k and the elongation of this string under the stress x:

$$PE = \frac{kx^2}{2}$$

1.9 Power

The definition of work and energy does not include anything about time spent for doing this work or losing/gaining this energy. But lifting the above mentioned object of weight 100 N to the height of 2 m could be done in 100 s or 1 s. Work done will be the same as well as energy gained by the object in this process.

If this task were done in 100 s, the average work done in this process would be 200 J/100s = 2 J/s. If the work were done in 1 s, the rate of work done would be 200 J/1s. = 200 J/s.

The work done per unit of time is called **power**. Power is formalized as

$$P = \frac{W}{t}$$

The units of power are W = J/s (watt). We all know this unit from everyday life. All electric bulbs are marked in accordance with the rate at which they convert electrical energy from a circuit into light and heat.

The electric company is using units of kilowatt-hour. These are units of energy, not power. This unit is an amount of energy consumed from the electric circuit when a device of power 1 kW worked for 1 hour. Now you can easily estimate which of your devices is the biggest "money-eater." Just check what power is shown in the description of your particular device.

Example 1.8. What work done corresponds to 1 kWh?

Answer:

1000 W x 3600 s = 3 600 000 J of work correspond to 1 kWh.

Example 1.9. What energy is stored in a spring of stiffness 2000 N/m stretched by 1 cm?

Answer:

$PE = kx^2/2 = 2000\ N/m \times 0.01\ m \times 0.01\ m/2 = 0.1\ J$

1.10 Applications to Sound

1.10.1 The Speed of Sound

While you are talking to your friend, standing close to you, it seems that sound arrives immediately as each word is spoken. But if your friend is standing at a distance of, say, 20, or better, 30 m from you, you begin to notice that sound arrives with a little delay. It means sound is traveling with finite speed.

Sound travels fast enough, and you should arrange special experiment to measure its speed. The speed of sound depends on temperature and could be written as:

$$V_s = 332\,\frac{m}{s} + 0.6 m/^0C \times T$$

where T is temperature in Celsius. This makes Vs = 344 m/s at 20°C and is equivalent to 1130 feet per second, or 770 mph. As you see, this speed is very high by everyday standards: no car can travel on a highway so fast. Only supersonic jets travel faster than sound.

Example 1.10. A rule of thumb that tells how close lightening has hit is "one mile for every five seconds before the thunder is heard." Justify, noting that the speed of light is so high that the time for light to travel is negligible compared to the time for sound. The temperature of the air is 20°C.

Answer:

The speed of sound at 20°C is

$$V_s = 332\,\frac{m}{s} + 0.6\,\frac{m}{^0C} \times 20\ ^0C = 344\,\frac{m}{s}$$

Time required for sound to cover 1 mi (1610 m):

$$\text{time} = \frac{\text{distance}}{\text{speed of sound}} = \frac{1610\ m/s}{344\ m/s} = 4s$$

For musical purposes it is important to mention that sound, whether high or low in pitch, travels in the air at the **same speed**. This is very fortunate for all music fans. Just imagine what would happen if treble notes, for instance, traveled faster than bass notes played by a musician at the same time. It means treble would reach our ears earlier than bass. Any chordal music would be totally destroyed!

As we see from the formula for the speed of sound, if the air temperature changes, so does the speed of sound. This has a simple explanation: with increasing of temperature, the air molecules move faster, the adjacent molecules collide more often, and they pass the disturbance of air faster from one region to another. In a warm room at 30°C, the speed of sound will be 350 m/s; that is nearly 2 percent faster than at 20°C. This is why wind instrument players blow gently through their instruments to warm them, filling them with air at needed temperature before performances. The pipe organs also have some problems because of the dependence of the speed of sound on temperature. Some reed pipes, such as Vox Humanas, tend to stay in tune regardless of the temperature; meanwhile flutes go off tune when the temperature varies.

1.10.2 Pressure and Sound Loudness

The loudness of sound is connected to a distance each molecule of air moves to each side relative to its undisturbed position. This is called **displacement amplitude**. For sound waves that are reasonably loud, this amplitude is pretty small—in the order of microns, which means a millionth of a meter or even less. Because we cannot see the molecules' motion and measurements of these displacements require special equipment, the concept of displacement amplitude is rarely used for the description of the loudness of sound wave.

It is far more convenient to use **pressure amplitude**. Just think about the process of the creation of sound: air molecules near the source start to vibrate, collide with adjacent molecules; they in turn collide with their neighbors, setting them into motion. This process results in areas of bigger (compressions) or smaller (rarefactions) pressure in comparison to average, atmospheric pressure without sound. Pressure amplitudes are also very small in comparison to 1 atm, but they could be easily measured.

The pressure amplitudes of sound waves at comfortable listening levels are from about 10^{-7} atm to 10^{-5} atm, or 0.01 N/m^2 to 1 N/m^2. So, the variation of pressure of 10^{-5} atm means that the pressure has the maximum value of 1.00001 atm in compressions and 0.99999 atm in rarefactions. If these variations are close to 1%, the sound of such amplitude will leave you deaf, setting the eardrums off.

Summary, Terms, Symbols, and Relations

Acoustics is an interdisciplinary science that studies sound.

For an object that travels at constant speed (v), distance (d) and speed are connected through a simple formula, where t is time:

$$\text{distance (d)} = \text{speed (v)} \times \text{time (t)}$$

Average acceleration (a) is defined as the rate of change of speed:

$$\text{acceleration (a)} = \frac{\text{change in speed}}{\text{time elapsed (t)}}$$

The second of Newton's laws combines the concepts of force (F) and acceleration. The acceleration of the object of mass m is proportional to the net force acting on the object:

$$F = m \times a$$

$$p = \frac{F}{A}$$

$$V_s = 332\,\frac{m}{s} + 0.6\text{m/°C} \times T$$

Questions and Exercises

1. What distance would sound travel in 1 minute if the temperature is 30°C? What distance would it be if the sound travels for one hour?

2. Estimate how long it would take for sound to travel from the front to the back wall in your auditorium. Do you hear any echo?

3. Estimate what pressure is exerted by a lady on the floor if the weight of the lady is 500 N and she is wearing sneakers. What would the pressure be if the same lady wore stiletto shoes with the area of the heel 1 cm^2?

4. You apply force of 5 N to the object of mass 1 kg that is sitting on a frictionless surface and it accelerates. What is the acceleration of this object? What acceleration will the object gain if the applied force is doubled?

6. If the pressure in rarefactions of some sound wave is 0.998 atm, what is the pressure in the compressions of this wave?

5. Suppose a sound wave has the pressure of 0.5 N/m^2. What is this amplitude expressed as a number of atmospheres?

7. Suppose you perform an experiment on measuring the speed of sound. Your friend is standing 100 m away from you and knocks a hammer against a wooden block. What time would it take for the sound to reach your ears?

8. A student lifts a weight of 100 N for 2 m and holds it in that position. How much energy is stored in this weight?

9. The weight of a car is 10000 N. Estimate the pressure exerted by each of the four tires on the ground if the area of contact of the tire is approximately 10 cm^2.

10. Which object is exerting bigger pressure on the support: a hard ball or a hard cube of the same weight?

Chapter 2

Simple Harmonic Motion

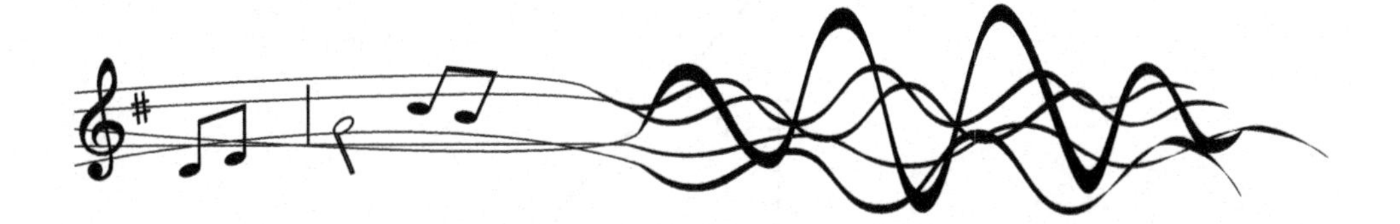

2.1 The Simplest Vibration

To consider something truly sophisticated, such as musical instruments, we should first look around for simple models. We know that to create sound, we should set some system (string, membrane, air inside of a woodwind) into vibration. Let's check around; what is the simplest possible system that demonstrates vibration? Sure, it is mass suspended on a string. At first glance, this system has nothing to do with music, but wait—we'll see how easily it could be turned into a guitar string.

The mass suspended on a spring, as shown in Fig. 2.1a, demonstrates our system at rest: the mass is at the equilibrium position, and the spring is somehow stretched, also in its equilibrium position. The force of gravity balances the elastic force from a spring. This elastic force depends on the elongation of our spring:

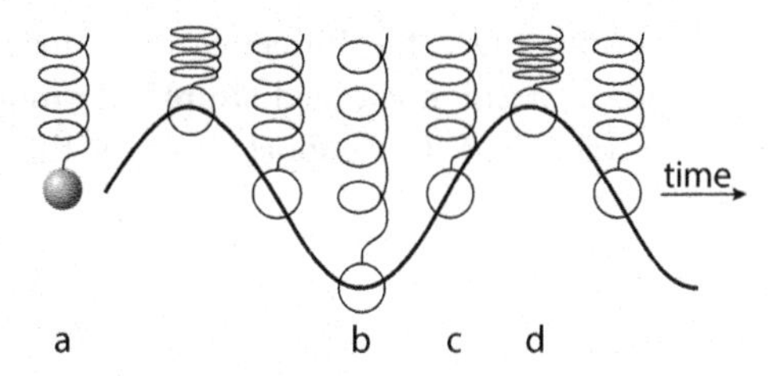

Fig. 2.1. The oscillations of a mass on spring with time.

$$F = -kx,$$

where x is elongation, and k is stiffness or spring constant. k depends on thickness of the wire as well as on the material our spring is made of. Please note the minus sign in the last formula: it demonstrates that elastic force is always directed against elongation (squeezing), trying to bring the suspended mass back to equilibrium point. This kind of force is called **restoring** force, and its presence is necessary for the existence of **any** vibration. Dependence of elastic force on elongation is shown in Fig. 2.2.

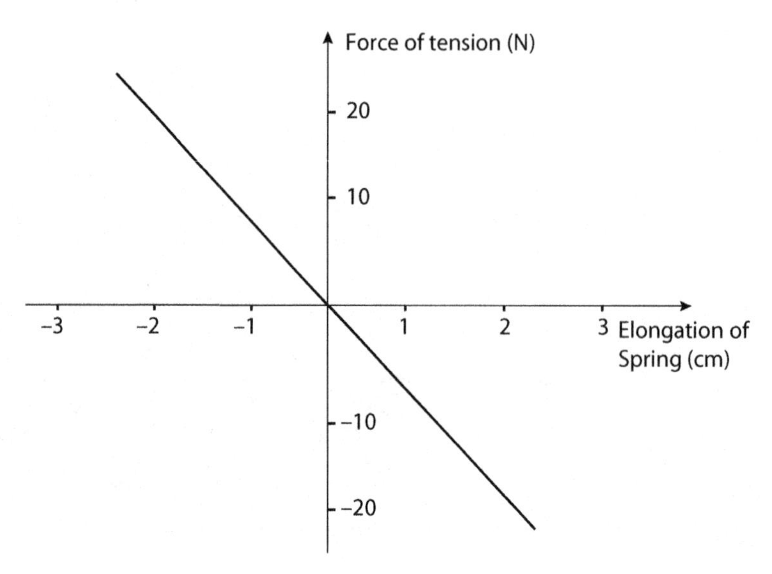

Fig. 2.2. Dependence of restoring (elastic) force on the elongation of a spring.

Now let's stretch the spring and release the mass. The restoring force, which is now bigger than the force of gravity, will bring the suspended mass back to the equilibrium position (see Fig. 2.1b). But the process will not stop at the equilibrium point: because of the nonzero speed of the mass, it will pass the equilibrium point (Fig. 2.1c), squeezing the spring (Fig. 2.1d). Restoring force is again directed toward the equilibrium point, and the process will repeat. If there were no friction or any other dissipative forces, such as air resistance, this process would repeat forever. We set the simplest system into vibration.

The motion under the influence of restoring force proportional to a displacement from equilibrium position is called **simple harmonic motion (SHM)**. Not many systems naturally demonstrate SHM. It is important for us that restoring forces of strings (guitar, violin, piano) are almost linear.

2.2 Period and Frequency of SHM

The time it takes to complete the full cycle of vibration is called the **period of oscillations** or **period of vibrations**. We will use the symbol T for this. Periodic dependence of position of mass on time is shown in Fig. 2.3.

Here is an important question that should be answered now because the consideration of different vibrating systems is pretty much the same, although systems could be different. The question is: what parameters will influence the period of our oscillations? And how?

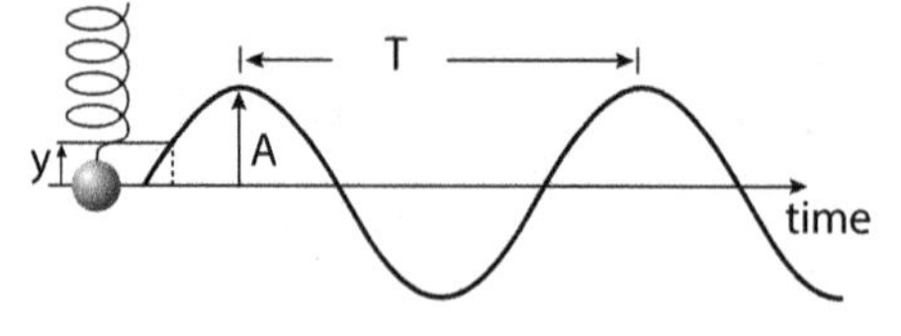

Fig. 2.3. Period of oscillations of a mass on spring.

At first glance, there are three important parameters:

1. **Mass** m suspended on a string. Intuitively we understand that using a bigger mass will provide a bigger period. A bigger mass will move more **slowly**, "lazily."

2. **Stiffness**, or spring constant k. The stiffer the spring, the faster a mass will move, so the period should get **shorter** with increase of stiffness.

3. The biggest displacement of a mass relative to the point of equilibrium is called **amplitude** of oscillations. It seems reasonable that this amplitude should also influence somehow the period of vibrations.

And here we are facing the huge advantage of SHM, the huge advantage of existence of restoring forces with linear dependence on displacement:

Period of SHM does NOT depend on amplitude of oscillations.

You could create oscillations with amplitude 1 cm, 2 cm, 3 cm … **if you have the same spring and the same mass, your system will demonstrate exactly the same period of oscillations.** We will discuss at the end of this chapter the crucial importance of this fact for the mere existence of music.

Period of oscillations could be written in formula:

$$T = 2\pi\sqrt{\frac{m}{k}}$$

Please note the square root dependence. The period of oscillations is **not** proportional to mass m. **For increasing** T **twice**, mass m should be increased **four** times at the same spring constant k. And **for decreasing** of T **twice,** you should increase spring constant k **four** times at the same mass.

For purposes of music it is much more convenient to use the concept of **frequency** instead of period.

Frequency f of oscillation is the number of oscillations per second, which is just a reciprocal of the period T of one vibration:

$$f = \frac{1}{T}$$

The unit of frequency is the hertz (Hz). One Hz is equal to one reciprocal second.

Example 2. A mass on a spring vibrates with frequency 5 Hz. How many oscillations per second do the prongs of this tuning fork perform? What is the period of oscillations?

Answer:

The period of vibrations is 1/f, or 1/5 Hz = 0.2 s. This system performs 5 oscillations per second.

2.3 Energy of the Vibrating System

In any vibrating system there is mutual exchange of two types of mechanical energy: potential and kinetic.

Let us consider this statement more in detail. When we stretch a spring (Fig. 2.1b), just before releasing a mass, is our system at rest? So kinetic energy is zero, but work done transfers into our system some amount of potential energy equal to

$$PE = \frac{kA^2}{2}$$

where A is an amplitude of our displacement. After releasing a mass, potential energy begins to decrease because of decreasing of displacement; meanwhile, kinetic energy, initially equal to zero, is increasing due to acceleration of mass. So, at some intermediate point, the energy of our system could be written as a sum of potential and kinetic:

$$PE + KE = \frac{kx^2}{2} + \frac{mv^2}{2}$$

While passing an equilibrium point, all the energy of our system is already stored in the form of kinetic energy:

$$KE = \frac{mv_0^2}{2}$$

where v_0 is the maximum speed of a mass. Note that potential energy at the equilibrium point is equal to zero because there is no elongation of a spring.

If there is no friction in the system considered, **the mechanical energy is conserved during the entire process**. This means that initial amount of potential energy stored in our system simply goes into the sum of potential and kinetic in intermediate points and goes into kinetic at the point of equilibrium. Thus, during the vibration, energy exchanges back and forth between kinetic and potential forms, but **the total energy stays the same**.

Dependence of potential and kinetic energies on time is shown in Fig. 2.4. These two energies are changing completely **out of phase**: when potential is a maximum, kinetic is equal to zero

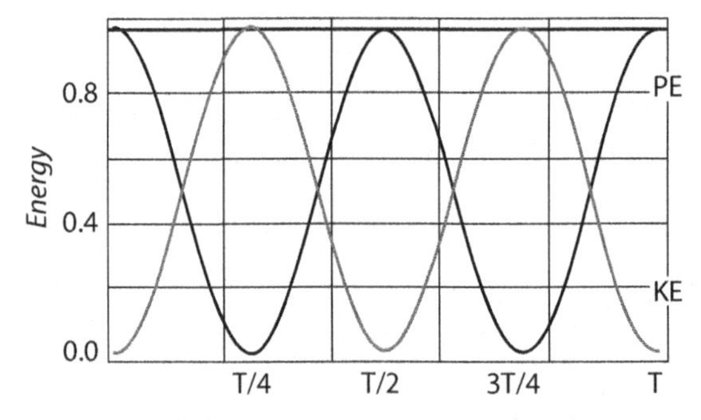

Fig. 2.4. Oscillations of potential (PE) and kinetic (KE) energy of a mass on spring with time. Note that PE and KE oscillate exactly out of phase.

and vice versa. Please note that the period of oscillations of energy is two times shorter than the period of oscillation of displacement.

Example 2.2. The maximum potential energy of the stretched string is 5 J. What is the maximum of kinetic energy? For what fraction of the period of oscillations will the system reach its maximum kinetic energy?

Answer:

The maximum of the kinetic energy is equal to the maximum of the potential energy, 5 J. As it follows from Fig. 2.4, the maximum of the kinetic energy is reached ¼ of period of oscillations after the maximum of the potential energy.

2.4 Another Vibrating System: Simple Pendulum

The simple pendulum is another system demonstrating SHM. It consists of a small bob of mass m and a massless (which means of a mass much smaller than the mass of the bob) rope or thread. This bob vibrates under the influence of a restoring force, which is a sum of two forces: gravity and tension of a thread (see Fig. 2.5).

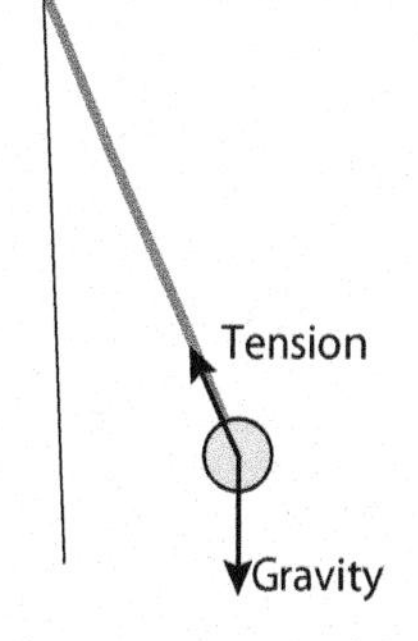

Fig. 2.6. Forces acting on a bob of a simple pendulum: gravity and tension.

For this system we also have a mutual exchange between potential energy (here it is the potential energy of gravity) and kinetic energy of the moving mass. The period of oscillations of a simple pendulum **does not depend on amplitude**, as well as for any system demonstrating SHM, and can be written as

$$T = 2\pi\sqrt{\frac{L}{g}}$$

where L is length of a thread and g is acceleration of the free fall due to gravity of our planet. Please note that acceleration of free fall is the same at any location on a surface of Earth: it is defined only by radius and mass of our planet. It means that if you will move pendulum without change of length to the surface of the Moon, it will oscillate with bigger periods because the free fall acceleration on the Moon is about six times less than that on the Earth.

Another important thing about a pendulum is that the period of oscillations does **not depend** on the mass of a bob. So, a bob of twice the mass 2m being suspended on a thread of the same length will demonstrate the same period of oscillation as a bob of mass m.

It could be useful to consider here a question, which many of us know since we were kids: Why does the swing move faster (that is, the period of oscillations decreases) when we are changing our position from sitting to standing up?

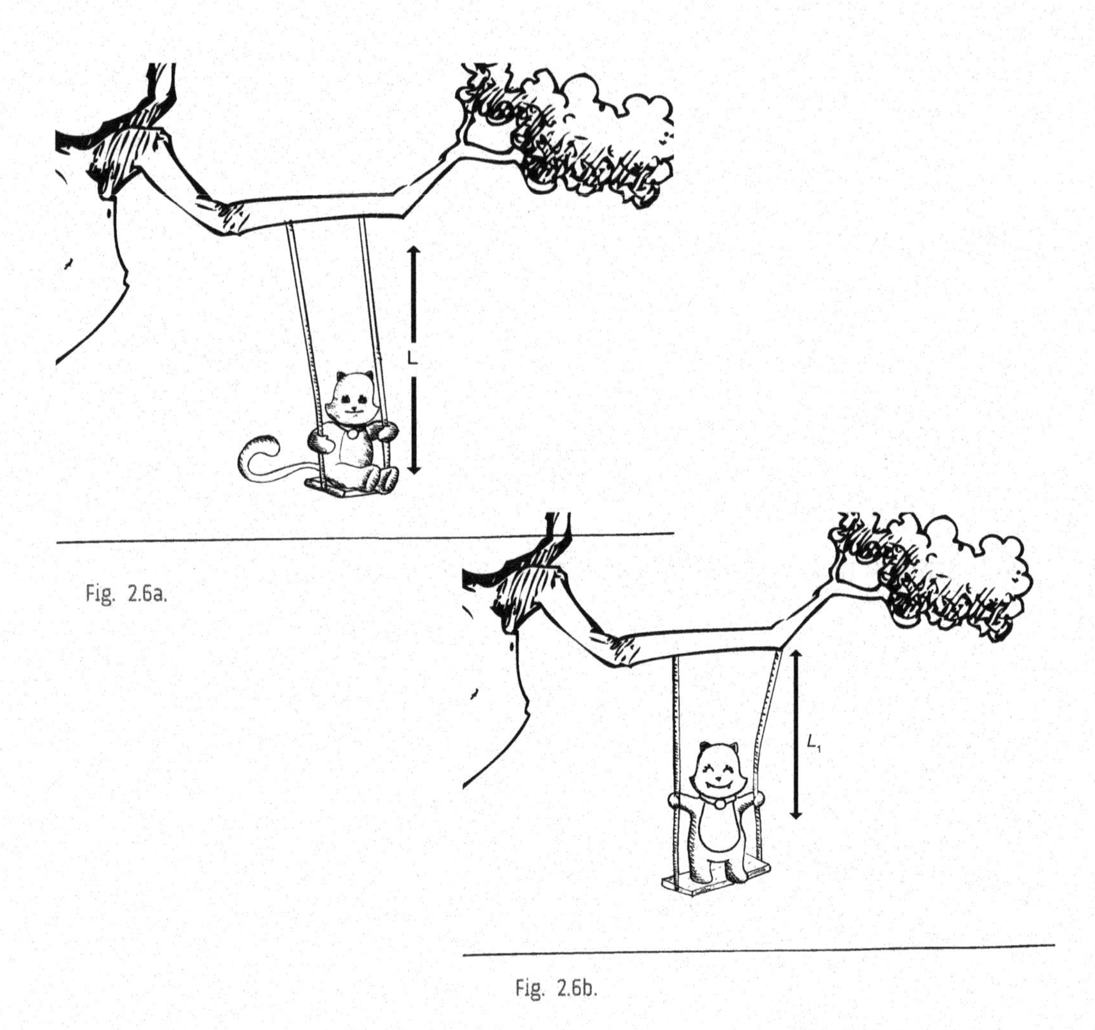

Fig. 2.6a.

Fig. 2.6b.

Fig 2.6a and 2.6b. The effective length of a pendulum decreases when we switch our position from sitting to standing up.

When we are sitting on a swing (Fig. 2.6a), the effective length of our pendulum is L. After we stand up, the effective length decreases because our center of gravity—the point of application of the force of gravity—moves up (Fig. 2.6b). Now the effective length of our pendulum-swing is $L_1<L$, and period of oscillations decreases.

Example 2.3. You increase the length of the pendulum 4 times. What would happen to the period (frequency) of oscillations?

Answer:

The period of oscillations of a pendulum relates to the length of the thread as a square root. So, if the length of pendulum is increased 4 times, the period of oscillations is increased 2 times. The frequency of oscillations, which is reciprocal to the period, is decreased 2 times.

Fig. 2.7. Resonance: When the frequency of the external force is close to the natural frequency of our system, the amplitude of oscillations increases.

2.5 Resonance

Consider a simple vibrating system: a child in a swing (Fig. 2.7). This swing has a natural frequency, which is determined by length and pull of gravity. With no friction involved, the displacement of this swing from its equilibrium point as a function of time is shown in Fig. 2.8a. Now we will give the swing a small push, called the driving force, exactly at the right time as it is shown in Fig. 2.8b with arrows. From everyday life we know that in this situation this swing, together with a child, will go higher and higher, which **means the amplitude of oscillations will drastically increase** (Fig. 2.8c).

The increasing amplitude of oscillations when the frequency of the driving force is close to the natural frequency of the system is called **resonance.**

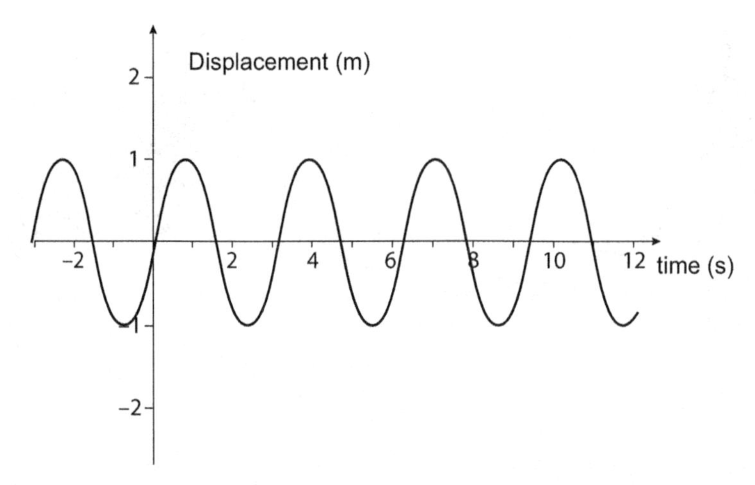

Fig. 2.8a. Displacement of the swing as a function of time, no external force applied.

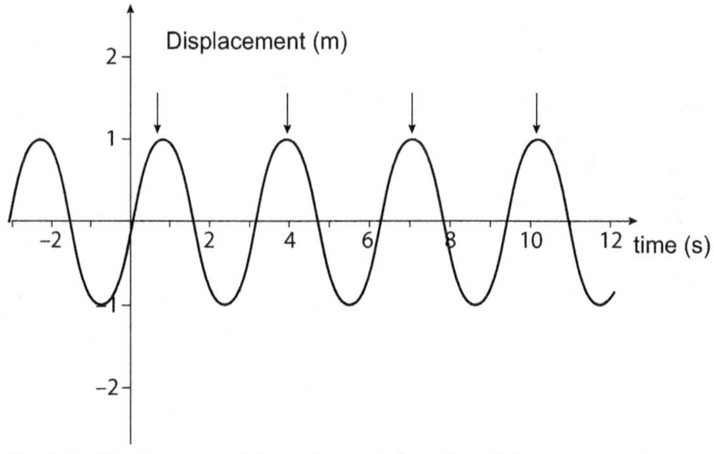

Fig. 2.8b. Displacement of the swing as a function of time, arrows show the moments of application of the external force.

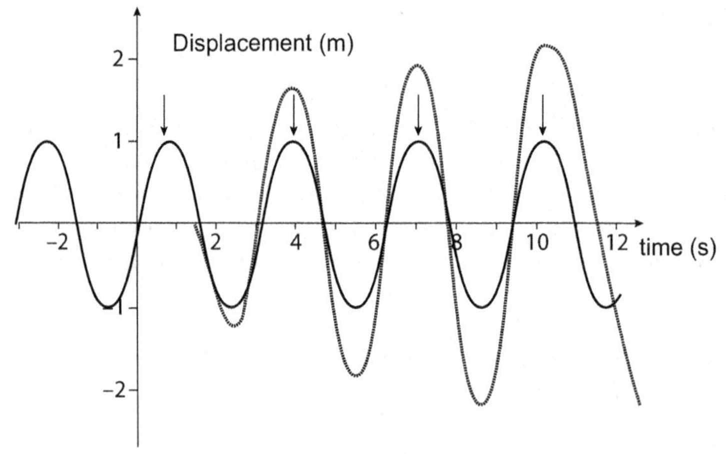

Fig. 2.8c. Increase of the amplitude of oscillations under influence of the external periodic force.

Resonance is possible in any system under the influence of a periodic driving force. If we give a small pull applied to a point of suspension to a mass on a spring (Fig. 2.9), it will also demonstrate resonance.

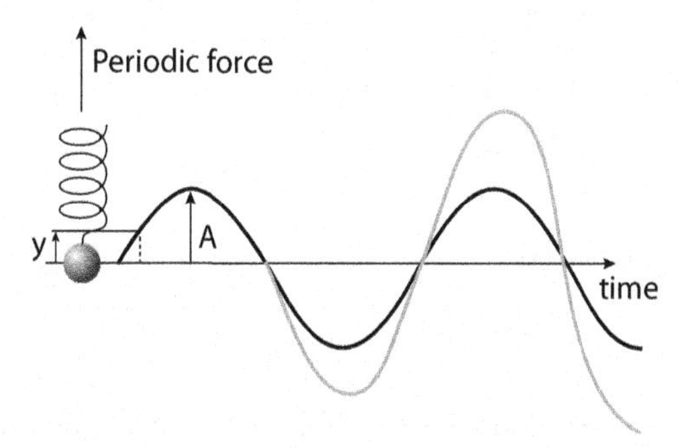

Fig. 2.9. Increase of the amplitude of oscillations of a mass on spring under influence of the external periodic force.

It is good to understand that this driving force need not be big; it could be rather small in magnitude, but how close the frequency of these pushes and pulls is to the natural frequency of a system has crucial importance for resonance appearance.

History shows a lot of examples when resonance played a destructive role. A lot of people think that the tradition of soldiers to break step while crossing a suspension bridge is nothing but superstition. But old times knew examples, when bridges have been destroyed by an army marching on them, when the cadence of soldiers' steps fell into the vicinity of resonating frequencies of these constructions.

But resonance could occur not only when the frequency of pushes and pulls is close to natural frequency. Let's come back to a child on a swing. What happens if we push the swing, not on each period, but on every other period as it is shown in Fig. 2.10 with arrows. The amplitude increases slower, but we again observe the resonance. This time resonance frequency is

$$f = \frac{f_0}{2}$$

and period

$$T = 2T_0$$

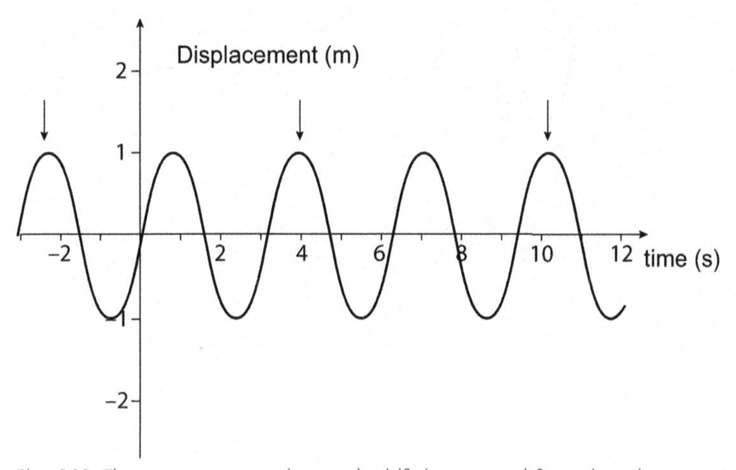

Fig. 2.10. The resonance can be reached if the external force has the period exactly two times bigger than natural period of oscillations of our system.

If the driving force is applied every third period of oscillations, we again have resonance. So, the general condition for the existence of resonance is:

$$f = \frac{f_0}{n}$$

and period

$$T = nT_0$$

where n is any integer. If there is no friction in a system, all these frequencies will provide the resonance.

2.6 Damping of Vibrations

Without friction, potential and kinetic energy oscillate in a vibrating system forever. But in real life we cannot get rid of friction. Any dissipative force (friction, resistance of air) is "stealing" part of the mechanical energy, turning it into another forms, such as heat, and the amplitude of oscillations gradually decreases each cycle of vibrations in the manner shown in Fig. 2.11. The dashed line showing the change of amplitude with time is called the decay curve, or just envelope. A vibrating system whose amplitude decreases is called **damped**, and the rate of decrease is the **damping** constant.

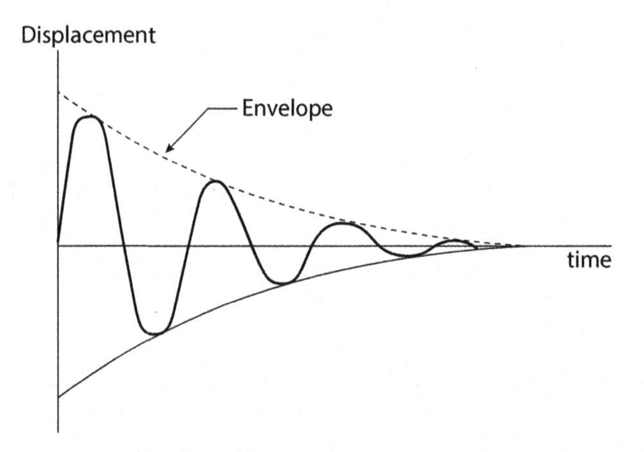

Fig. 2.11. Damped oscillations: decreasing of amplitude with time.

2.7 Applications to the Sound

2.7.1 Vibration of the Guitar String

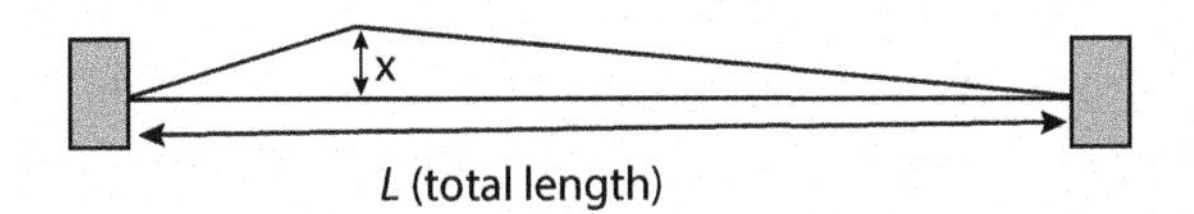

Fig. 2.12. Guitar string displaced from equilibrium position.

A guitar string, when displaced a small distance x (Fig. 2.12) from its equilibrium point, has potential energy

$$PE = \frac{2F_t}{L}x^2$$

where Ft is the force of tension in a string. Once released the potential energy of tension is changed into kinetic energy, and again energy changes between kinetic and potential.

The analogy between a string and a mass on a spring is not very clear: the role of mass is played by the mass of the spring per unit length (m/L) and the role of spring constant k by the force of tension. Everybody who played a guitar knows that thick strings with big ratio m/L demonstrate lower frequency in comparison to thin strings. Also we know that the increasing of tension leads to the increasing of frequency.

Frequency of strings in musical instruments does not depend on amplitude, exactly as we have seen for mass-spring system. It does not matter how hard you hit a piano key, say A^4. Soft or loud, which means small amplitude or big amplitude, it will still oscillate with a frequency of A^4.

2.7.2 A Helmholtz Resonator

A Helmholtz resonator is just a cavity with a narrow opening (Fig. 2.13). This type of air vibrator was named after H. von Helmholtz, who used it to study the creation of musical sounds. The cavity of these resonators can have different shapes and sizes. The simplest example of such a device is just an empty plastic bottle. If you blow air across the top of it, you hear a clear musical tone of low frequency.

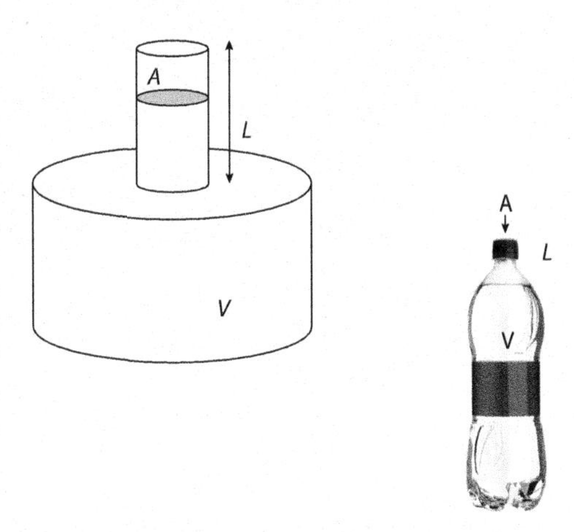

Fig. 2.13. Helmholtz resonator. The usual plastic bottle can serve as a Helmholtz resonator.

At first glance there are no springs and masses in this system. But let's analyze Fig. 2.13 in detail. The air in the narrow neck oscillates back and forth, playing the role of a mass on a spring. Meanwhile, air inside the cavity of volume V is springy and can be considered as the spring. The frequency of vibration is such in a resonator

$$f = \frac{V_s}{2\pi}\sqrt{\frac{A}{Vl}}$$

where A is the area of the neck, shown in Fig. 2.13; l is its length; V is the volume of the cavity; and v_s is the speed of sound.

Note that the bigger the area of the neck A, the higher the frequency that is demonstrated for a resonator at a given length and volume.

2.7.3 Breaking a Glass by Resonance

Everybody has heard the legends that opera singers can break wineglasses with their voices. They say that the great bass opera singer Feodor Shalyapin shattered small glass ornaments on the chandelier during his performance. Is it possible? Perhaps. This process does not have any contradictions with Physics.

If the performer sings a note at precisely the frequency of the natural vibration of a wineglass, it will excite resonance in the glass. This note could be easily determined by tapping the glass lightly on the rim. But the amplitude of a singing voice alone, without amplifiers, is small, and the time during which a performer can hold this note is not long enough. Feodor Shalyapin had an extremely strong voice and perfect breath technique, so maybe legends about him are not just legends.

But still, it is possible to break glass with sound if this sound is enhanced with amplifiers and loudspeakers. All videos that could be found online show the breaking of wineglasses with sound created by a sound generator, not by a human voice.

2.7.4 Sympathetic Vibrations

This kind of vibration occurs when an initially passive vibrating system responds to external vibration close to the natural frequency of the passive system.

Let us consider a classical example: two tuning forks, one of which is mounted on a wooden box (Fig. 2.14). If the other one is struck and then brought to the opening of the box, then muted, the un-struck fork mounted on a box will be heard. If these two forks are not identical, no sound will be produced. In this situation **sympathetic vibration** is excited close to the natural frequency of the initially passive system (mounted tuning fork).

If we will just take a vibrating tuning fork and press its base against a wood plate or tabletop, we will hear the significant increase of sound amplitude. The difference between the situation considered above is that any other tuning fork will also demonstrate louder sound after pressing it against a big

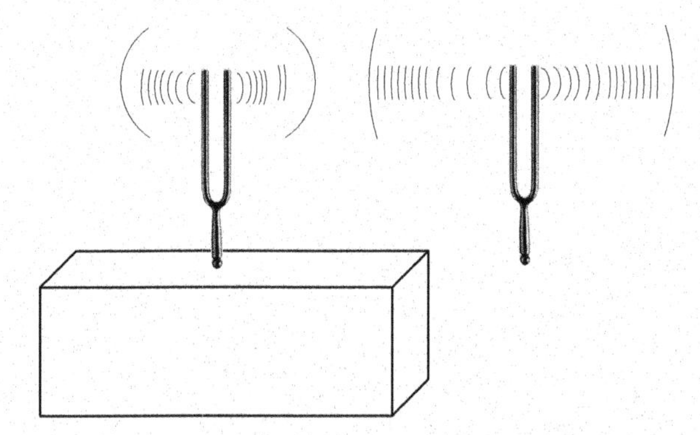

Fig. 2.14. Sympathetic vibration is excited close to the natural frequency of the initially passive system.

surface. Why? Any surface, such as a tabletop, has not one but many resonating frequencies, so the frequency of a tuning fork will fall with great probability in the vicinity of one of them.

The phenomenon of **sympathetic vibration**, which is also nothing but a resonance, is widely used in construction of musical instruments. Let's take, for example, strings. A vibrating string alone displaces very little air as it vibrates and, as a result, radiates a very small amount of sound. But if we connect it to a box, the sympathetic vibration of a box will move much bigger masses of air because much a larger surface is set into vibration.

All acoustical string instruments depend mostly on sympathetic vibrations of the wood sounding box for radiation or their sound. Most of the sound radiation in these instruments is produced by sympathetic vibrations of the top plate, which is set into vibration from the vibrating string through the bridge. The top plate has many resonances distributed over the playing range. Sympathetic vibration of the wood also sets the air inside the instrument into vibration. These are so called air resonances, extremely important for the quality of an instrument. The sound enhanced by sympathetic vibrations of the air radiates though the sound hole of a guitar or f-holes of a violin. String instruments will be discussed in Chapters 11 and 12.

Pianos and other keyboard instruments such as harpsichords and spinets have large soundboards with a lot of resonances, closely spaced through the range of an instrument. These soundboards are set into sympathetic vibrations through bridges as in guitars and violins. We will discuss pianos in Chapter 11.

2.7.5 Damping of the Sound

Resonance is a very useful phenomenon, without which music simply would not exist. But excess resonance could be crucial to a performance. For instance, pianos have soft dampers made of felt to damp the vibrating strings. Without these dampers, any note of the piano, which has really good resonating soundboards, will last for tens of seconds, making the playing of fast staccato a total mess.

If bigger times of decay are needed, the right pedal (so called "loud" pedal, which is not exactly a valid name) is pushed down. This pedal rules the dampers, lifting them from the strings. Each note lasts several times longer than with right pedal released. More on this topic will be discussed in Chapter 11.

Summary, Terms, Symbols, and Relations

Simple Harmonic Motion (SHM): mass on a string, simple pendulum.

Period (frequency) of SHM does not depend on the amplitude.

Period of a mass on a string:

$$T = 2\pi\sqrt{\frac{m}{k}}$$

Period of a simple pendulum:

$$T = 2\pi\sqrt{\frac{L}{g}}$$

Frequency of the Helmholtz resonator:

$$f = \frac{V_s}{2\pi}\sqrt{\frac{A}{Vl}}$$

Resonance is the increasing of the amplitude of the vibrations when the frequency of the driving force is close to the natural frequency of the system.

Questions and Exercises

1. A mass on a spring vibrates with the frequency 7 Hz. What is the period of vibrations of this mass? How many vibrations per second does this mass perform?

2. A mass on a spring is vibrating with the frequency 5 Hz and amplitude 1 cm. We stopped this motion and started the new one, with amplitude 2 cm. How does the frequency of vibrations change?

3. Let us imagine a grandfather pendulum clock that keeps perfect time on Earth. Discuss what would happen to the period of its vibrations if this clock were brought to the Moon. Hint: the gravity of the Moon is weaker than the gravity of Earth.

4. Consider two springs, A and B. The spring constant of A is twice as big as the spring constant of B. They are loaded with identical masses and set into vibration. What is the ratio of the periods of these two springs? Which spring demonstrates a bigger frequency of oscillations?

6. Two pendulums, A and B, have threads that are the same length but the mass of bob A is twice as large as the mass of the bob B. Which pendulum demonstrates a higher frequency of vibrations?

5. Estimate the natural frequency of a 0.5 L plastic water bottle if the temperature of the surrounding air is 20°C.

7. Discuss what would happen to the period of oscillations of a swing if you change your position from sitting on the seat to standing on it.

8. Some system vibrates with a natural frequency of 100 Hz. What frequencies of external force will produce resonance?

9. Consider two springs, A and B, which have the same spring constant. Spring A is loaded with mass 2 kg and spring B with mass 1 kg, and then they are set into vibration. Which spring will demonstrate a higher frequency of oscillations? What is the ratio of these frequencies?

10. What happens to a frequency of a Helmholtz resonator if the temperature of the surrounding air changes from 10°C to 30°C? Compare the change of frequencies with the figures from Appendix B. Will you be able to hear the difference?

CHAPTER 3

Waves

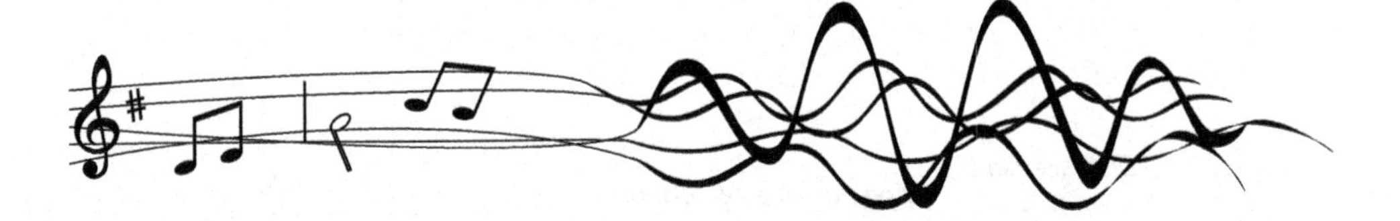

3.1 Creation of Waves

In the previous chapter we considered simple harmonic oscillation and came to the conclusion that the simple example of mass on a spring gives us a lot of facts to discuss about vibrating musical systems, such as strings and air columns. A string or an air column inside a woodwind could be called a source of disturbance, which, after creation, starts to travel through the elastic medium, usually through the air.

Let us perform a mental experiment because it is not so easy to set up in reality. In Fig 3.1a the mass and spring shown are connected to an elastic cord of infinite length. In the real world we simply do not have elastic cords that are infinitely long. When the mass on a spring is set into oscillation, it also pulls on the end of the elastic cord secured to it, creating **waves** on this cord. In Figs. 3.1b and c we show the momentary "snapshots" of this wave.

Fig. 3.1a. Mass on spring is at rest, so attached cord is undisturbed.

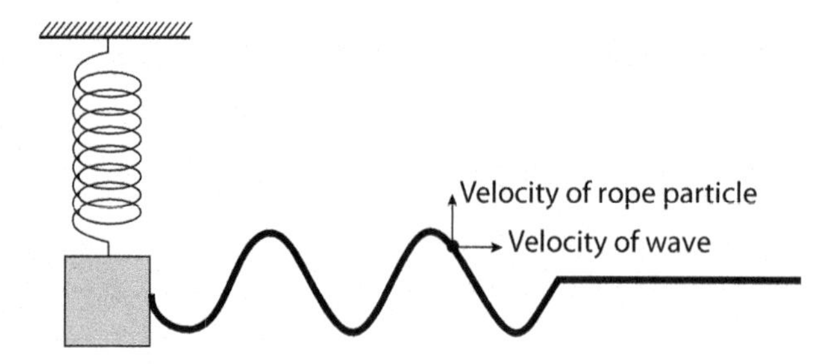

Fig. 3.1b. The oscillations of a mass on spring produce the waves on the cord. Velocity of the wave as well as velocity of the element of the cord is shown.

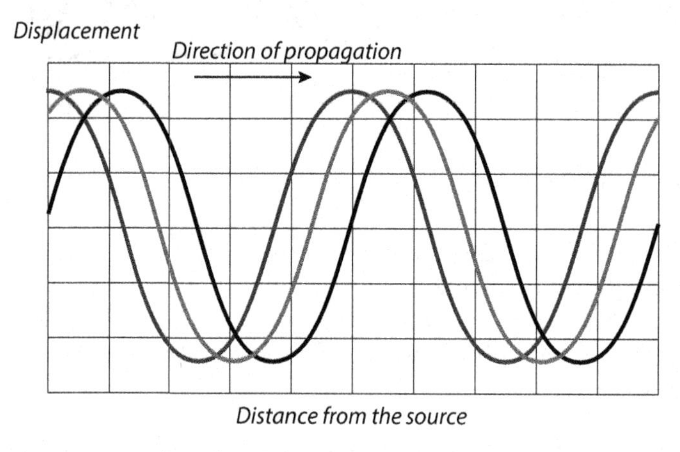

Fig. 3.1c. Several profiles of the wave shown as time progresses.

Because a mass on a spring performs perfect SHM, waves are strictly periodic both in **time** (with period or frequency corresponding to ones of oscillating mass) and in **space**. We need to consider now two types of graphs: dependence on time at a given point of space (Fig. 3.2) and dependence on distance from the source at a given moment of time (Fig 3.3).

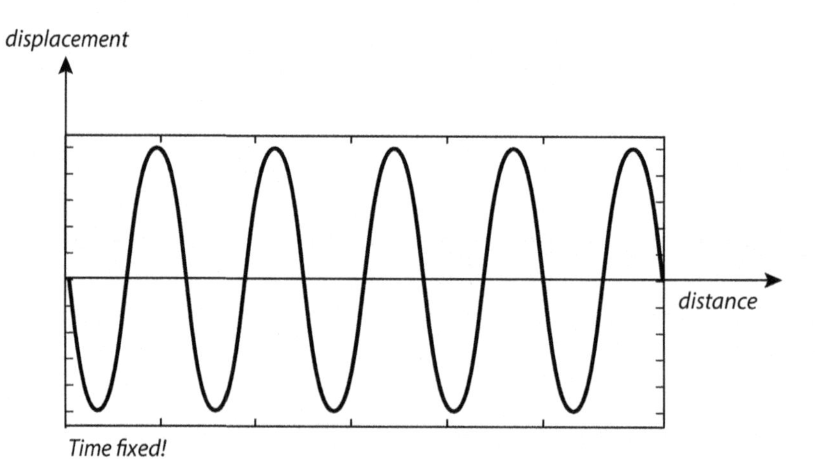

Fig. 3.2. The snapshot of a traveling wave. Time is fixed, the wave is shown as a function of a distance.

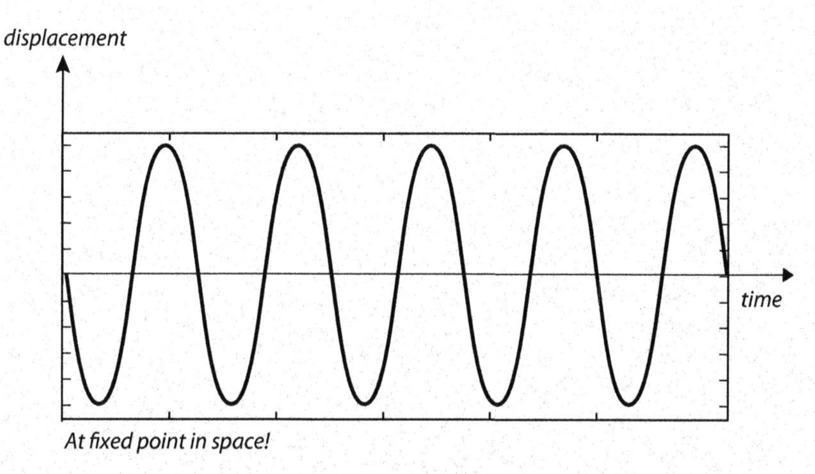

Fig. 3.3. Change of the displacement at given (fixed) point in space as a function of time.
Adapted from Physics: Principles with Applications by Douglas C. Giancoli, 2004 Pearson Education, Inc.

The vibrating source can perform not only SHM. If we take the end of a cord and just give it a small fast displacement back and forth, the wave created will have a form of a short pulse traveling along the cord (Fig. 3.4).

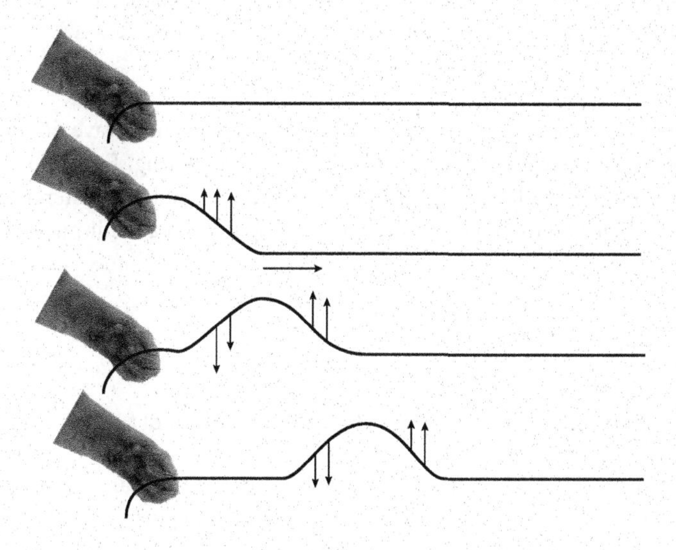

Fig. 3.4. Creation of a simple pulse wave on a cord.
Copyright © 2013 by Depositphotos Inc./oksana tkachuk.

3.2 Motion of Particles in the Wave

Almost all waves, except for electromagnetic (light, radio, TV), should have a medium to travel. In the example considered above, the medium for created wave was a cord. Remove the cord, and the wave will disappear. The particles of elastic medium (air, water, cord) are set into motion while the wave is going through them.

Let's take a rope and make a mark with a pen at some point on it (Fig. 3.5). This mark will help us to understand how the actual element of a medium is moving when a wave is passing through it. If we excite waves on this rope, we see that the particle of a rope **does not move** along with a wave. It is just performing back-and-forth oscillation, staying in the same place.

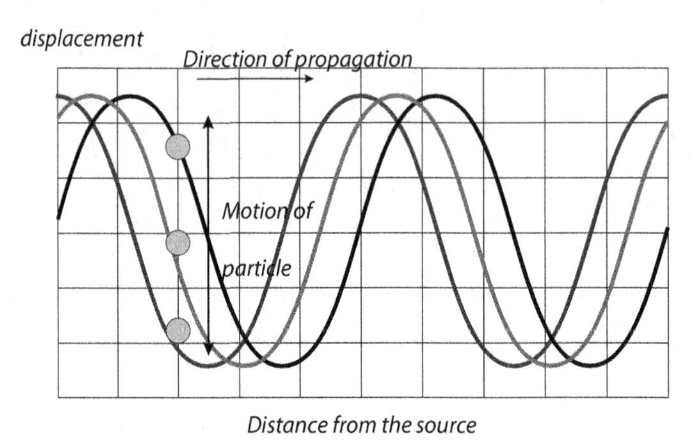

Fig. 3.6. The motion of a particle in a wave. Note: the particle is performing oscillatory motion but not moving along with the wave.

This is a very important property of a wave. Particles of air in a sound wave do not move all the way from your lungs to the ear of your friend while you are talking. Each molecule oscillates back and forth around the equilibrium position. Thus, there should be a difference between the speed of a single particle and the speed of a wave traveling though a medium as a whole.

3.3 Types of Waves

The type of wave is completely defined by the character of motion of a particle in it.

In a **transversal** wave (Fig. 3.6) a particle is oscillating perpendicular to the direction of the propagation of a wave. Examples of such waves are electromagnetic (light, heat, radio, TV, X-rays), waves on the surface of water, destructive transversal waves of an earthquake, and waves on a cord as considered in our mental experiment.

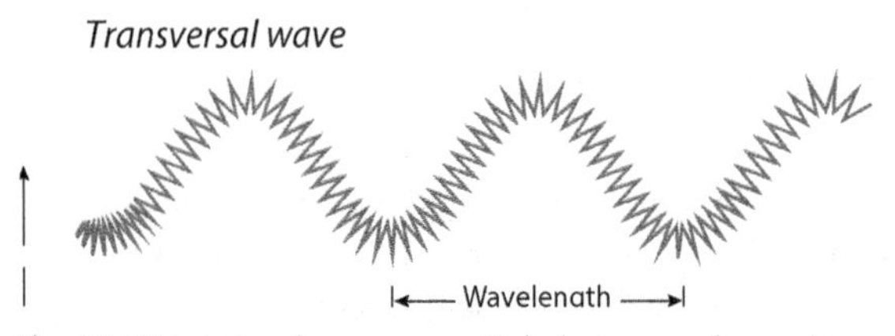

Fig. 3.5. Transversal wave: a particle in a wave is moving perpendicular to the direction of propagation.

In a **longitudinal** wave (Fig. 3.7) a particle is oscillating parallel to the direction of the propagation of a wave. The areas of elastic medium are compressed and expanded under the influence of a wave. Longitudinal waves mostly exist in gases and liquids. Examples of longitudinal waves: sound, longitudinal waves of earthquakes, wave on a Slinky.

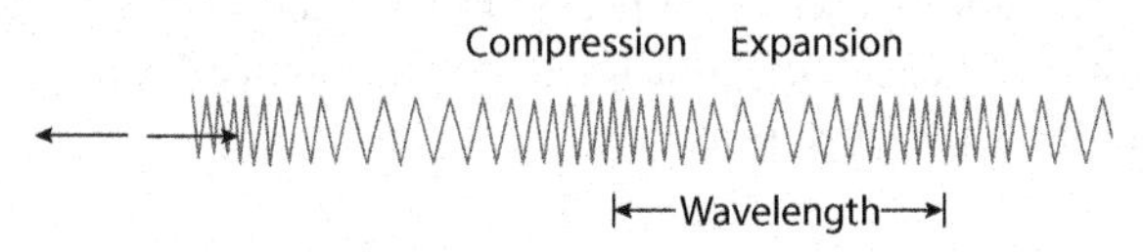

Fig. 3.7.

3.4 Periodic Waves

The source vibrating periodically with time creates waves that consist of repeating patterns in time and space (Figs. 3.8a and b). Note that the source can perform not only SHM but any periodic motion. Such waves are called **periodic waves.**

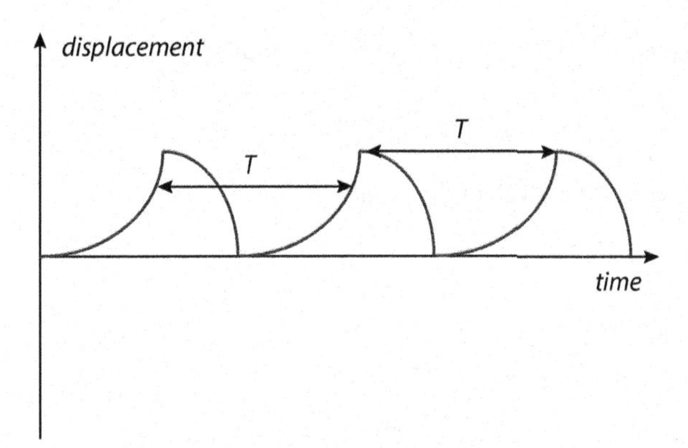

Fig. 3.8a. Periodic, but not sinusoidal, wave as a function of time. The period of the wave is shown.

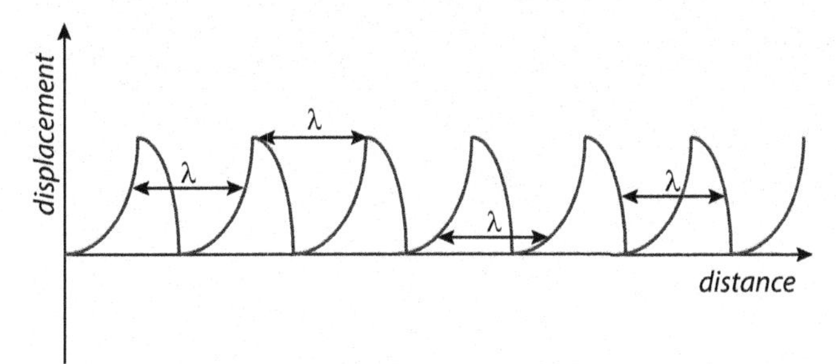

Fig. 3.8b. The snapshot of a periodic, but not sinusoidal, wave; the wavelength is shown.

Periodic waves have two characteristics: period (or, reciprocal to it, frequency) in time, and period of repletion in space, which is called wavelength. The symbol traditionally used for the last is the Greek letter λ. Wavelength is measured in units of length, such as meters, centimeters, etc. Wavelengths are also shown in Figs. 3.6 and 3.7 for transversal and longitudinal waves. Note that a wavelength can be measured, not just from maximum to maximum but also from any point to the next **identical** point.

Because a periodic wave repeats itself, each period in time and each wavelength in space, it is obvious that during the time equal to T, our wave travels distance λ. So, the speed of any wave can be written though these parameters as:

$$v = \frac{\lambda}{T}\lambda f \quad 3.1$$

So, at given speed of wave (such as sound or electromagnetic), the bigger the wavelength, the smaller the frequency.

Example 3.1. What is the speed of periodic wave with frequency 10 Hz and wavelength 2 m? What would the speed be if the frequency is increased by two, keeping the wavelength the same?

Answer: Following

$$v = \lambda f = 10Hz \cdot 2m = 20\, m/s$$

if the frequency is increased by two and wavelength stays the same:

$$v = 20Hz \cdot 2m = 40\, m/s$$

Example 3.2. What is the wavelength of red light?

Answer:

The speed of light is 3×10^8 m/s, the frequency of red light is approximately 4×10^{14} Hz. As a result, the wavelength of red light is

$$\lambda = \frac{v}{f} = \frac{3 \cdot 10^8\, m/s}{4 \cdot 10^{14}\, Hz} = 7.5 \cdot 10^{-7} m$$

Example 3.3. What wavelength corresponds to the sound wave (temperature 20°C) produced by note A^4?

Answer:

The frequency of A^4 is 440 Hz, the speed of sound at given temperature

$v = 332\frac{m}{s} + 0.6\frac{m}{s \cdot {}^\circ C} \cdot 20^\circ C = 344\frac{m}{s}$ As a result, the wavelength

$$\lambda = \frac{v}{f} = \frac{344\frac{m}{s}}{440\, Hz} = 0.78\, m$$

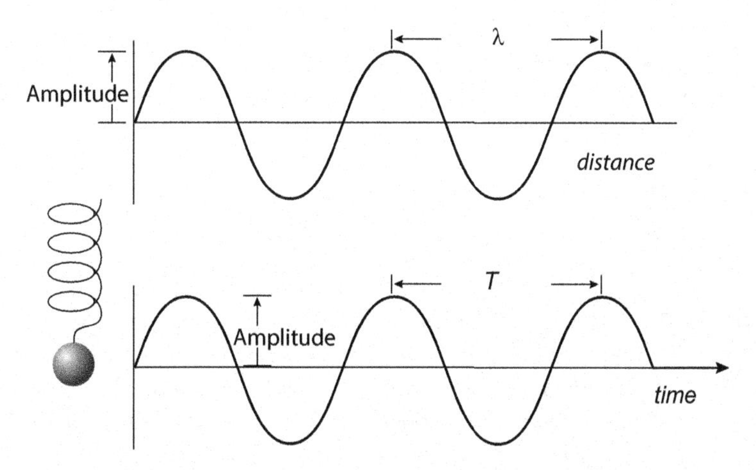

Fig. 3.9a. The snapshot of the sinusoidal wave created by SHM.
Fig. 3.9b. Sinusoidal wave created by SHM as a function of time.

3.5 Sinusoidal Waves

If a vibrating source performs SHM, each particle of a medium in the wave created also undergoes SHM. This kind of wave, as shown in Fig. 3.9, is called **sinusoidal**, or just **sine, wave.**

The full importance of sine waves in the Science of Sound will become apparent when we discuss, in Chapter 9, how complex motions such as in Fig. 3.8 can be represented as combinations of several simple sinusoidal waves.

3.6 Influence of a Medium on the Speed of Elastic Waves

When a vibrating source creates oscillation and a wave, it actually gives only the frequency of oscillations. A mass on spring does not care what would happen to an infinitely long cord far, far away; maybe it will go under the surface of some lake. A violin string, vibrating at frequency 440 Hz (A^4), does not care if this sound will strike the water; the string will continue to vibrate at the same 440 Hz. But the speed of the wave is changing, changing the wavelength as a result. Wavelength is fully determined by the property of the medium—in other words, it depends on how hard it is for a wave to go through this medium (Fig. 3.10).

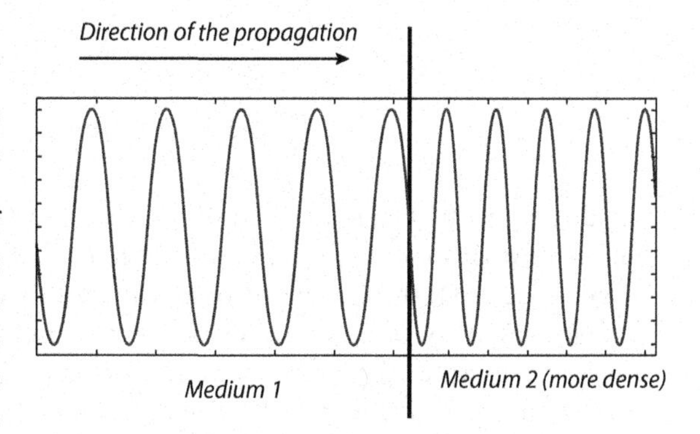

Fig. 3.10. The propagation of the wave through the boundary between two media.

Speed on any wave in a medium depends on two important factors: elastic properties (which means how **resistive** a medium is to propagating a wave) and density of the medium, which is just mass per unit volume.

The general formula for speed of any elastic wave looks like

$$V = \sqrt{\frac{B}{m/V}} \qquad (3.2)$$

where B is elastic (bulk) modulus of a medium, which is an analogy of stiffness, m mass, and V volume of the vibrating medium. The bigger B is, the more resistive the medium is for wave propagation. From this formula, it is obvious that the speed of sound—the elastic wave we are interested in—will strongly depend on properties of the medium and differ for air and, for instance, water.

If we have a medium that is arranged along one long line (such as a rope or a string), the speed of the elastic wave in it is defined by

$$V = \sqrt{\frac{F_t}{m/L}}$$

where F_t is the force of tension of a string, m mass, and L length of the string.

Example 3.4. The rope of mass 0.1 kg, length of 1 m is under tension of 40 N. What is the speed of the elastic wave created in this string?

Answer:

$$V = \sqrt{\frac{F_t}{m/L}} = \sqrt{\frac{40\ N}{\left(0.1\ kg/1m\right)}} = 20\ Hz$$

3.7 Applications to Sound

3.7.1 Creation of Sound

Sound is an elastic longitudinal wave, which is created by oscillations of a density of a medium or by oscillations of pressure in the medium. The sound wave looks schematically as it is shown in Fig. 3.11: the membrane or a drumhead is vibrating, setting into vibration adjacent air molecules. Areas of compression and expansions are changing each other. In compressions the average distance between molecules is less than in areas of expansion. Wavelength could be measured from, say, the middle of a compression to the middle of the next compression.

Compressions and expansions shown in Fig. 3.11 are highly exaggerated. In reality, as it was already discussed in Chapter 1, the variations of pressure and, as a result, the density of the air are very small. For a very loud sound, such as one from a jet aircraft 40 m away, the variations of pressure are about 200 Pa, or 0.002 atm, and this already exceeds the threshold of pain, which is 64 Pa, or 0.00064 atm. A rock concert, several meters from amplifiers, creates pressure variations of 20 Pa, or 0.0002 atm, and this is a threshold of discomfort for the average person. Usual conversation has only 0.02 Pa of pressure variations. Finally, the threshold of hearing, the softest possible sound that the average person can hear, has 0.00002 Pa of pressure variations.

Although the pressure varies only slightly in a sound wave, the range of these variations is pretty wide. The difference in pressure between a rock concert and the threshold of hearing is one million times. How our ear is able to accept signals of such different amplitudes we will discuss in Chapter 7.

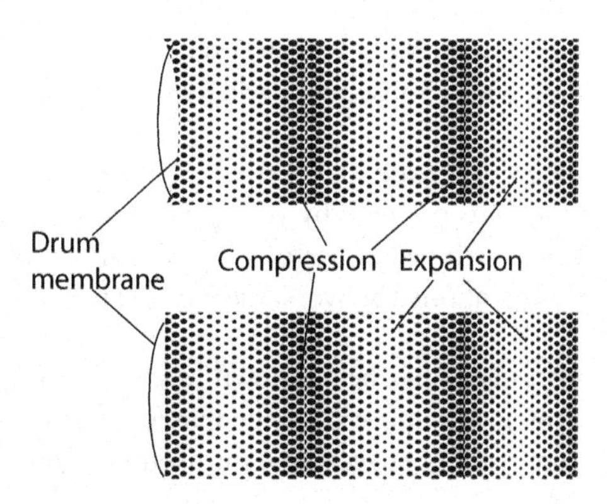

Fig. 3.11. The sound wave created by a drum membrane. The areas of expansion and compression are shown.

3.7.2 Audible Frequencies and Wavelengths

When we hear sound, it means that our eardrums are set into a vibration of proper frequency. The audible frequencies of sound, meaning the frequencies at which we actually hear and perceive as the sound, vary by individual, but for the average person they are between 17 Hz and 17,000 Hz.

Frequency is responsible for perception of **pitch**. The human hearing apparatus is pretty sophisticated, as we will see in Chapter 7, but so far it is enough for us to understand that the bigger the frequency of oscillations of the eardrum, the higher the pitch we hear.

The limit of hearing for a particular person depends on different factors, such as, for instance, age. With age, the upper limit of hearing gets lower.

Note that these limits of audibility are properties of human hearing only. The limits of hearing in dogs are between 40 Hz and 60,000 Hz, which is much greater than that for humans. This property is used in so-called dog whistles, which create sound of frequencies well above the upper limit for humans but are audible to dogs.

Bats have an even higher upper limit of hearing: 120,000 Hz. Bats need very sensitive hearing because of their lack of visual ability and use the upper frequencies for echolocation.

The hearing range of mice shifts drastically into the range of higher frequencies. They can perceive sound from 1 kHz to 90 kHz.

But the widest hearing range is demonstrated in marine mammals. For instance, bottlenose dolphins hear sounds from 0.25 Hz to 150 kHz. This species uses high frequencies for echolocation and low frequencies for social interaction.

The human body reacts on any frequencies of elastic waves, but only at between 17 Hz and 17,000 Hz do we perceive the signal as sound. The waves of higher frequencies, which are called

ultrasound, we practically do not feel, because a human is too big and too heavy to oscillate in sympathy. Ultrasound is widely used for medical purposes for the study of internal organs.

Below 17 Hz there is an area called **infrasound**. The frequencies of it are so low that you can feel vibrations not only by your ear but also by the whole body. Infrasound often results naturally from earthquakes, waterfalls, tsunami, and other phenomena on large geometric scales. As it was mentioned above, some animals use infrasound to communicate over long distances.

Infrasound has been known to cause feelings of fear in humans. Because it is not perceived consciously, it can make people feel that some supernatural events are going on around them. Some studies suggest that infrasound signals are even responsible for ghost sightings.

Frequencies and wavelengths are connected through 3.1, so small frequencies of the sound correspond to big wavelengths and vice versa. The audible range of wavelengths is about from 2 cm (approximately 1 inch) to 20 m.

3.7.3 The Speed of Sound in Different Media

Sound needs media to travel. **There is no sound in a vacuum.** Only one kind of wave, which we encounter every day, can travel though empty space: the electromagnetic one. Both heat and light, as well as radio waves, reach us every moment from our star, Sun, after covering huge distances of interplanetary vacuum. From the point of view of sound, the space above Earth's atmosphere is the most silent place possible. In some sci-fi movies you can see a character falling out of a spaceship and receding from it, screaming out loud. This is absolutely impossible and can be considered only as some kind of artistic approach. With the exception of electromagnetic waves, the only way to transfer a signal in empty space is to use some available elastic medium. For instance, if you are outside a spaceship, you can knock on some metallic surface. Sound travels inside the metal will reach the ears of somebody who is inside the spaceship.

As we see from Fig. 3.2, the speed of sound in different media depends on the density of this medium and its stiffness. Both of these factors matter. For instance, water is approximately 1000 times denser than air, but the stiffness of water is greater than the stiffness of air—much more than 1000 times. Water is almost incompressible, like a very stiff spring. As a result, the speed of sound in water is approximately 1,600 m/s, almost 5 times faster than in air.

The speed of sound in solids is much higher than in the air. For instance, in aluminum, sound travels at 6420 m/s and in steel at 6100 m/s. Sitting on the rails, you feel the train approaching first with the parts of your body which are in contact with rails, and only after this do you hear it through the air.

The speed of sound in the gas helium, with which balloons are filled, is 965 m/s. Helium is a very light gas; however, its stiffness is approximately the same as that of the air. In another noble gas, xenon, the speed of sound is only 170 m/s. All these numbers are given for 20°C.

The slowest speed of sound is demonstrated in different kinds of stretchable material whose stiffness is pretty small. For instance, the sound in rubber travels with the speed of 40–150 m/s.

The speed of sound in the air, given in Chapter 1, is defined for the sea level densities of dry air. There are two factors influencing the speed of sound with altitude: temperature is getting significantly lower and density of air is getting thinner. The speed of sound around commercial jets cruising at altitude 10,000–20,000 m is only 290 m/s, with the temperature of -60°C.

Example 3.5. How long would it take for a sound to travel the length of your lecture hall at 20°C? What would this time be if the sound travels under water?

Answer:

The length of my lecture hall is approximately 35 m, the speed of sound at given temperature is 344 m/s. The time required for sound to cover this distance

$$t = \frac{35\,m}{344\,m/s} \cong 0.1s$$

The sound would cover such distance under water approximately five times faster at 0.02 s.

3.7.4 The Influence of Waveforms

The simplest possible periodic wave is the sinusoidal wave. The pure sine wave is of little musical interest. Sine wave is created, for instance, while whistling softly. It has definite musical pitch, but its **timbre** is not very pleasant.

We already know that changes of pressure in a sound wave are responsible for loudness of sound. This dependence is not very simple and will be discussed in detail in Chapter 7.

Frequency is responsible for pitch.

But there is also another characteristic of sound. We could easily hear the difference between the note played by a violin and the same note hit on a piano. But the note is the same; it has the same pitch and may have the same loudness.

Periodic waves can not only be sine waves; moreover all sound waves created by musical instruments, including the human voice, are **not** sine waves at all. Fig. 3.12 shows the sound waves created by different instruments playing the same note. They all have the same period of oscillation but absolutely different waveforms. As we will see in Chapter 9, all these waves are different combinations of sine waves with frequencies corresponding to multiples of the frequency of the playing note.

Waveform is what defines the **timbre** of a particular instrument.

Some instruments have a different timbre when played by a different performer. For instance, the beginner playing violin creates a sound wave with a form close to that shown in Fig. 3.13a; meanwhile, the more experienced performer extracts from this violin a sound of a form close to that shown in Fig. 3.13b.

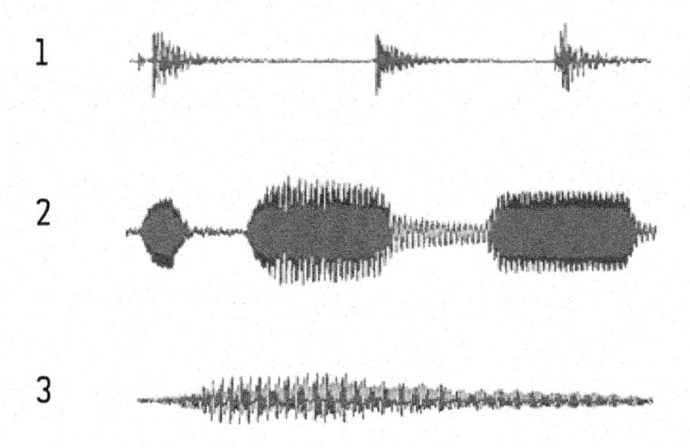

Fig. 3.12. Waveforms produced by: 1) four notes from a tabla (Indian drum); 2) three notes from a French horn; 3) one note from a flute.

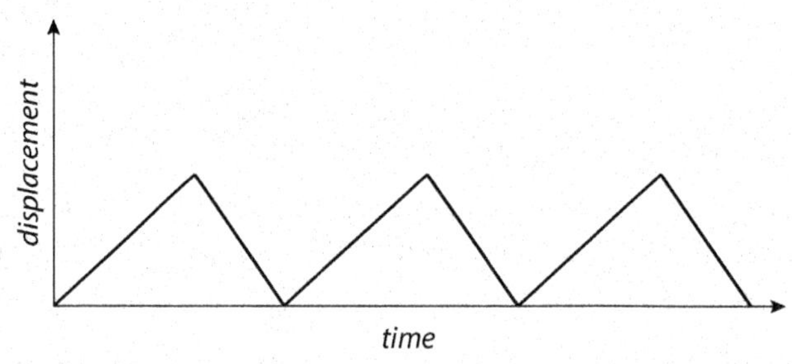

Fig. 3.13a. The simplified view of the waveform created by a beginner playing violin.

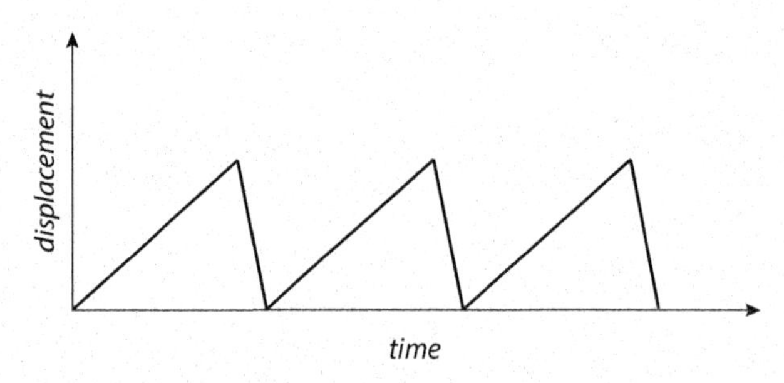

Fig. 3.13b. The simplified view of a the waveform created by an experienced performer playing violin.

Summary, Terms, Symbols, and Relations

A periodic wave repeats itself both in space and time.

If the source of the wave vibrates following SHM, the wave created by this source is the sinusoidal wave.

There are two types of waves: transversal (like light) and longitudinal (like sound).

The speed of periodic wave, its wavelength, and its frequency are related through

$$v = \frac{\lambda}{T} = \lambda f$$

There is no sound in a vacuum. Sound needs a medium to travel.

The audible frequencies of sound for an average person are between 17 Hz and 17,000 Hz and audible wavelengths are between 2 cm and 20 m.

Questions and Exercises

1. The sound of the note A_3 on a piano has the frequency 220 Hz. Does this sound travel faster or slower than the sound created when the note A_4 (440 Hz) is hit?

2. A violin string vibrates at frequency 1720 Hz. What is the wavelength of the sound produced in the surrounding air at temperature 20°C?

3. What is the wavelength of the sound with frequency 200 Hz compared to the wavelength of the sound of 300 Hz?

4. If an organ pipe is built to produce sound of wavelength 3 m, what frequency will this sound have? (Assume temperature 20°C.)

6. Suppose you are watching water waves pass a tide gauge at the end of the pier. At their highest point, the crests reach a mark labeled 4.0 m and at the lowest point, the 3.0 m mark. What is the amplitude of these waves?

5. Calculate the wavelengths of the extreme bass and uppermost keys of a piano. Use for reference figures from Appendix B.

7. The string of mass 5 g and length 1 m is under tension of 100 N. What is the frequency of oscillations of this string?

8. The speed of light is 3×10^8 m/s. The wavelength of red light is 650 nm. Estimate the frequency of red light waves.

9. The tuba produces the sound wave of wavelength 2 m in the air with temperature 20°C. What is the frequency of the produced sound?

10. Estimate how many wavelengths will fit the length of your auditorium when the note A_4 (440 Hz) is played.

CHAPTER 4
Families of Musical Instruments

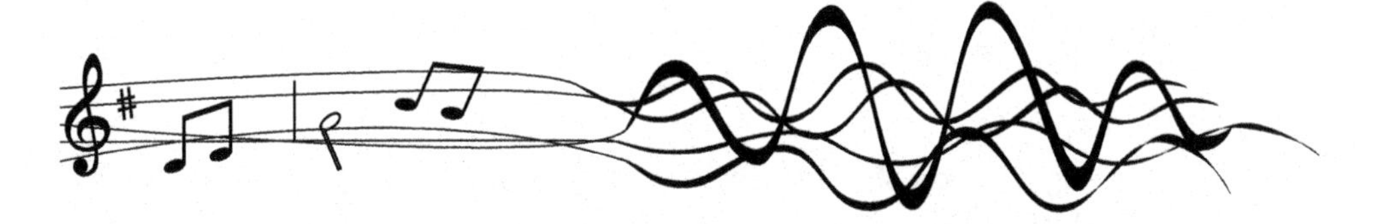

4.1 Different Sound Sources

From previous chapters we know that for the creation of sound, we need **a source vibrating at frequencies between** 17 and 17000 Hz. The world around us contains numerous vibrating systems, which create all types of sounds. It is hard to imagine a situation of absolute silence. But even if you sit quietly in a room with walls that do not transmit any sounds from outside, you still will hear a very soft sound created by the fluids of your own body. Let us think about the classifying of sounds into different categories.

First, we should consider **natural** and **artificial** sounds. Nature produces many sounds around us, but only a few of them are musically useful. The vast majority of these sounds, although generally pleasant, like rustling leaves in a fall forest, are just noise, which means these sounds are unorganized. Most music contains sounds that are produced by strictly controlled processes, which we want to understand and identify.

Second, sounds can be considered as being **original** or **reproduced**. Sound can be recorded in the memory of your computer or any CD writer, stored, and played many times. The most important characteristic of this process is how close the stored replica is to the original source. This is mostly the problem of the quality of your recording device. In this book we are interested mostly in original sounds.

Another fundamental difference is between **transient** and **steady** sound. Transient sound occurs for a short period of time, usually much less than a second. It appears when a source is set into vibration by a brief interaction, such as the clack sound from two wooden sticks. Such sound dies very quickly after being created (Fig. 4.1) and usually does not have definite musical

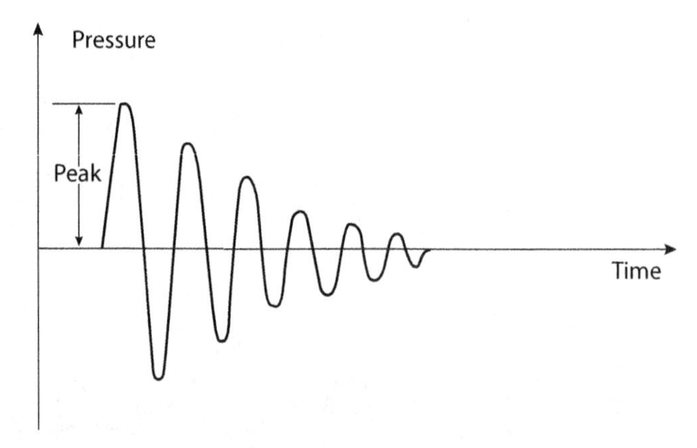

Fig. 4.1. Decay of pressure oscillations in a typical transient sound wave.

pitch. For steady sounds we need steady input of energy; for instance, a bow moving on violin strings, which maintains the permanent vibration of the sound source.

The musical sounds are usually classified by the nature of vibrating source and how this source has been set into vibration. It means we should consider families of musical instruments. The problem here is that during its history, humanity created many different kinds of musical instruments, so we should subdivide any family into smaller groups. For instance, a piano is a string instrument, but its strings are struck by hammers, so in some books it is referred to as percussion. The recorder and the clarinet are both wind instruments, but they have different resonators and different ways of excitation of sound. The celesta has hammers striking metal bars and should belong to the percussion family, but the character of a resonator (a pipe for each pitch) makes it related to the woodwinds. So, we consider some general features of operating basic instruments in an orchestra.

4.2 Percussion Family

What happens when we strike one hard object against the other? We hear the sound, usually transient, which dies really fast. Why does it happen? While in contact, these two objects exert force on each other, which causes them to deform. After the contact was lost, the surfaces start to vibrate, also setting into vibration the areas adjacent to the point of actual striking. These vibrations come to a stop sooner or later at the position of equilibrium. How soon the vibration will stop depends on the elastic properties of our surfaces.

What we have just described is the easiest way to understand the operation of any percussion instrument, such as a drumhead or xylophone. Any rigid surface is not absolutely rigid and is capable of carrying the initial disturbance within itself (Fig. 4.2). This disturbance inside of our rigid material can exist in the form of both longitudinal and transversal waves. This vibration disturbs the air molecules, which create disturbance in the air, already traveling to our ears in the form of the longitudinal sound waves.

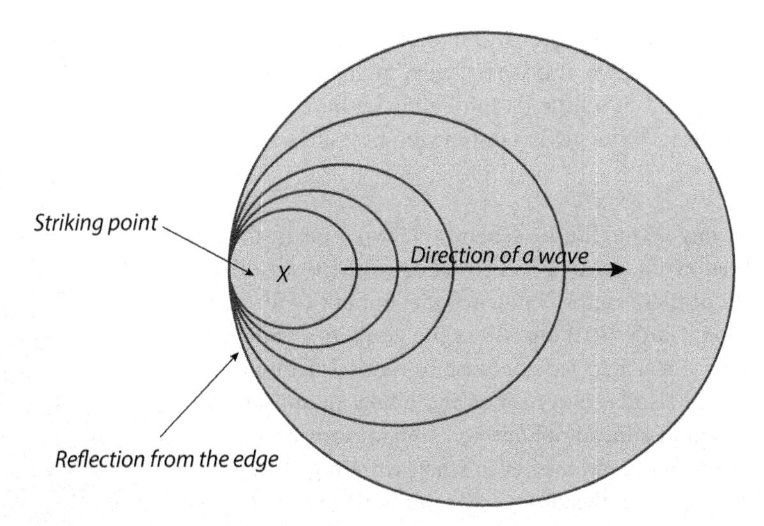

Fig. 4.2. Propagation of the sound wave on a surface of a drumhead.

For most percussion instruments, sound does not produce a clear sensation of pitch, although it does give you some impression as to whether the pitch is high or low. This means that surrounding air vibrates at a high or low frequency. Instead of some definite frequency corresponding to one particular note of a piano, percussion demonstrates the whole permanent band of frequencies. These frequencies are determined by both the size of the vibrating object and the properties of the material, such as stiffness and density. These two, as we know from Chapter 3, define the speed of sound in a medium.

For example, if we strike two bars of different lengths, we create sounds of different frequencies because for a long bar it will take longer for a disturbance to travel back and forth; as a result, the longer bar produces the sound for a longer period or with a lower frequency. If you take two bars that are identical in shape and size but made of different materials, the stiffer bar will produce a higher frequency. In Chapter 10, we will consider formulas containing size and stiffness of bars for precise analysis.

Some percussion instruments, such as metal bells of specially designed shapes or a hung drum, do produce sound of definite clear pitch (Figs. 4.3a and 4.3b). In Chapter 10 we will discuss the influence of the form of these instruments on their pitch and timbre.

The sound of a percussion instrument is always transient; it fades away in a fraction of a second (for a wooden block or xylophone), or it could last for a couple of seconds, as with a heavy church bell. The main reason for the fading of a sound is the transmission of vibrating energy into motion of air molecules. The vibrating surface pushes on surrounding air, losing energy itself. This explains why sometimes the vibration of small objects, such as a tuning fork, lasts longer than the vibration of

Fig. 4.3a. Typical hung drum.

big drumheads: a tuning fork sets into motion smaller amounts of air, and as a result, energy is transferred away at smaller rates. The case of big church bells is a little bit more difficult because the bell itself, due to its special form, has not only a vibrating surface but also a resonator.

The sound, varying in amplitude, depends on two aspects: the amplitude of vibrations of the area and the surface of the area that is vibrating. Both of these factors influence the amount of air that is set into vibration. Harder striking will create vibrations with larger amplitude; when it starts to fade because of natural damping of a process, the sound also becomes softer and finally disappears. The small vibrating bar of a metallophone and a big drumhead will create a sound of different loudness, even when vibrating at the same amplitude, because a drumhead sets into vibration a bigger amount of air.

Fig. 4.3b. Typical hand bell.

Another way of losing energy during vibration is from different kinds of internal friction. The vibrating object is heating while flexing due to finite elasticity. Most of the energy of xylophone bars is dissipated because of this particular mechanism.

We can divide the standard percussion instruments into subfamilies based on the material of the vibrating object.

Membranophones, or simply drums, have a uniform membrane fastened on a circular loop. Membranes were traditionally made of animal skin, but now artificial materials, like different plastics, are more common mostly because the properties of artificial materials depend on such factors as humidity and temperature in much less measure than a natural one does. Membranophones have different mechanisms for the regulation of a desired tension of a membrane. For instance, tympani have a pedal mechanism with which a performer can easily switch the tension while going to another pitch.

Fig. 4.4. Example of a metallophone: glockenspiel.

Metallophones can have either definite or indefinite pitch. The high stiffness of metals usually results in a low internal friction, so there is usually a small loss of energy to heat. Vibrations of metallophones usually die away slowly in comparison to vibrations of a skin (plastic) membrane or wooden block.

Metallophones of definite pitch include vibrophones and bells. The keyboard metallophone celesta looks pretty much like a piano, but the hammers strike not string but metal plates suspended over wooden pipes. The celesta is used, although not very often, in modern orchestra. The best-known music piece that uses the celesta is Tchaikovsky's "Dance of the Sugar Plum Fairy" from *The Nutcracker.*

Another important metallophone, a glockenspiel, consists of a set of tuned keys arranged in the fashion of the keyboard of a piano (Fig. 4.4). One of the most widely known classical pieces using the glockenspiel is Mozart's "Die Zauberfleute."

The **xylophone** consists of wooden bars struck by mallets. Each bar, like in a celesta or glockenspiel, is tuned to a pitch of the musical scale. This instrument has ancient origins. According

Fig. 4.5. Example of a keyboard: harpsichord.

to different sources, it originated in Southeast Asia. The first use of the xylophone in European orchestra was in "Danse Macabre" by Saint-Saens to imitate the sounds of rattling bones.

4.3 String Instruments

A stretched string can be considered as a very long and thin metallic bar. But as we will discuss in Chapters 11 and 12, a stretched flexible string has different properties, which are very important for musical properties. First of all, a string always provides definite pitch.

Instruments in which the vibrating system is a string are called string instruments. But again we come to the necessity of subdividing this big family into smaller subclasses. The timbre of an instrument strongly depends on how our string is set into vibration. We can strike it with a felt-covered hammer (piano, clavichord, tsymbaly), we can bow it (violin, viola, cello), we can pluck it (guitar, sitar, balalaika, harp). They are just a few examples of each—in reality, the list of string instruments is as long as a mile because almost each country or nation in the world has its own string instrument. There are also string instruments that are played another way than mentioned above; for instance, the aeolian harp is a string instrument "played" by the wind.

The keyboard instruments, such as piano and harpsichord (Fig. 4.5), have one or more strings for each note to be played. The harp, which is a plucked instrument, also has one string per note.

Fig. 4.6. The change of pitch on the guitar is reached by the decreasing of an effective length of the string.

Another class of string instruments has only a few strings. We obtain several notes from each of them using different portions of its length. We can shorten the effective length of a string by pressing a string at one point (Fig. 4.6). On some instruments, guitars, for instance, there are thin strips called **frets** glued on fingerboards, defining the points against which to press the

Fig. 4.7. A single string does not move substantial amounts of air; as a result, the sound created is very soft.

strings. The shorter the portion of a string participating in the vibration, the higher the pitch. For a string this dependence is very simple: frequency is indirectly proportional to the length of a string. This means if we decrease the length of the string by half, the pitch will go twice as high. Some instruments, such as violins, do not have frets, so the player can extract from violin any pitch not only those corresponding to the note of a piano. This gives some more freedom, but also produces a lot of difficulties for a performer: to play proper notes, the string should be pressed by "feeling of the ear" only. Everybody knows that playing the violin requires a very good musical ear.

A string itself moves very small amounts of air, creating a very soft as well as perfectly tuned sound. The pushing of air by a string is extremely inefficient (Fig. 4.7). To create a loud sound we should mount this string on a bigger surface, which, vibrating in sympathy, starts to move big amounts of air. All acoustic instruments—instruments that are not connected to any electric amplifiers—have mechanical **resonators**, wooden boxes of sophisticated forms (Fig. 4.8). The vibration of a string is transmitted to this box or other surface, setting it also into vibration. This vibration is responsible for almost all sound

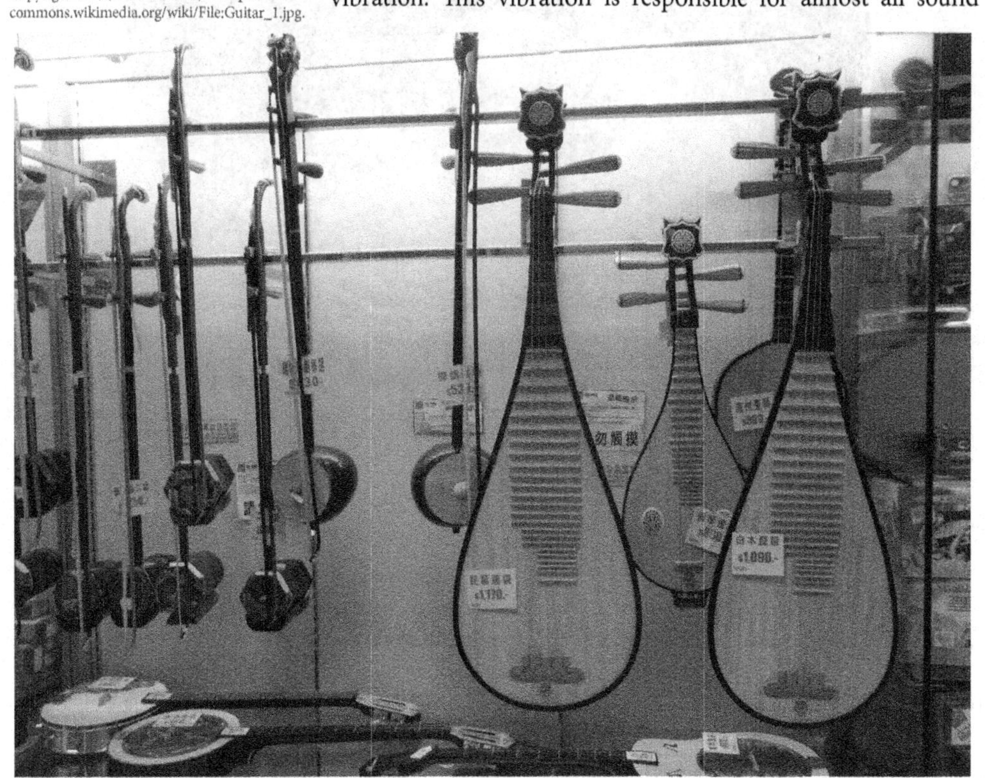

Fig. 4.8. Different resonators of string instruments.

emitted by an instrument. Note that this process does not create additional energy: relatively high amplitude of oscillations of a string is transmitted into the much smaller in amplitude vibrations of a resonator, which is much heavier and bigger than string itself. Sound, amplified by a resonator, dies faster because the radiation of the soundboard drains the energy from the string at a much higher rate.

We can create the vibration of a string maintained over a long time by bowing it. We will discuss the precise mechanism of this in Chapter 12.

4.4 Wind Instruments

Another way to deliver a steady energy supply to maintain a steady vibration is by using an airstream. The wind instrument also contains a resonator, usually a pipe, in which a column of air is set into vibration by the player blowing through some sort of mouthpiece placed at one

Fig. 4.9. The brass family: the role of mouthpiece is played by the configuration of performer's lips.

end of the resonator.

The wind instruments are usually subdivided into three classes, according to the three different ways to blow the air into the tube: brass, reeds, and edgetones.

In the **brass** family, the role of a mouthpiece is played by the lips of a performer (Fig. 4.9). The player controls the tension in the lips, so they vibrate under the influence of the airflow going through them. This is the main difference between brass and other wind instruments. Most of the members of this family are really made of brass, but not all of them. The wooden

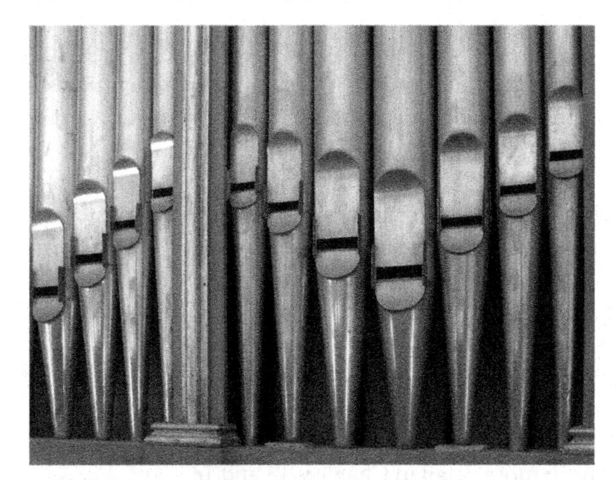

Fig. 4.10. Organ flue pipes.

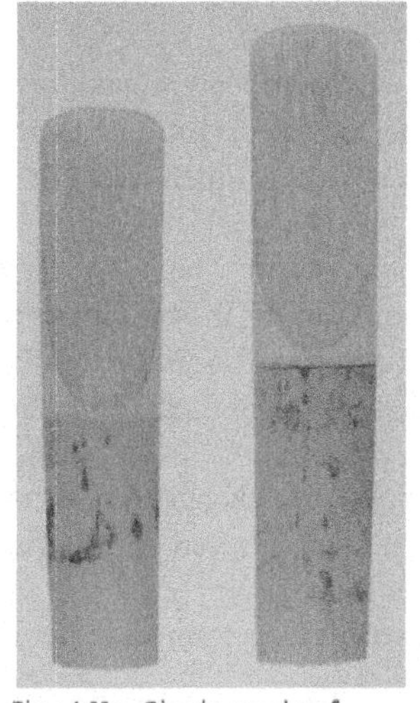

Fig. 4.11a. Single reeds of woodwinds.

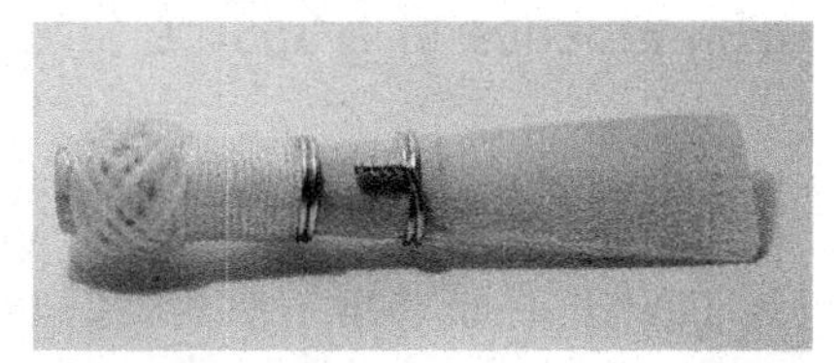

Fig. 4.11b. Double reed of woodwinds.

cornet and the serpent are both made of wood or even plastic tubing but belong to the brass family because the vibration is caused by the player's lips.

In **edgetones**, such as flutes and recorders, the vibration is created by blowing air against the edge. The edge is breaking an initially uniform airstream into pulsating vortexes that propagate outward. An edgetone has a definite pitch, often with some accompanying hissing sound, like a whistle. If we connect this edgetone to a resonator of a proper size and shape, this resonator cuts from a sound of an edgetone the vibrations of a particular frequency.

Most organ pipes (Fig. 4.10), called organ flue pipes, work this way. Air under pressure is driven down the flue against the sharp lip, causing the air in the pipe to vibrate with a frequency determined by the pipe length.

In **reed** instruments sound is produced by vibrating a thin strip of a material, biting against the edge of an instrument. This strip, which is called a reed, is usually made from cane, but in modern instruments synthetic materials are also used. Reeds can be single (Fig. 4.11a) like in a saxophone, or double (4.11b) like in an oboe or bassoon. In the case of the crumhorn and bagpipes, a reed cap that contains an airway is mounted above the reed (4.11c), with a player blowing only to keep pressure in that chamber above the atmospheric.

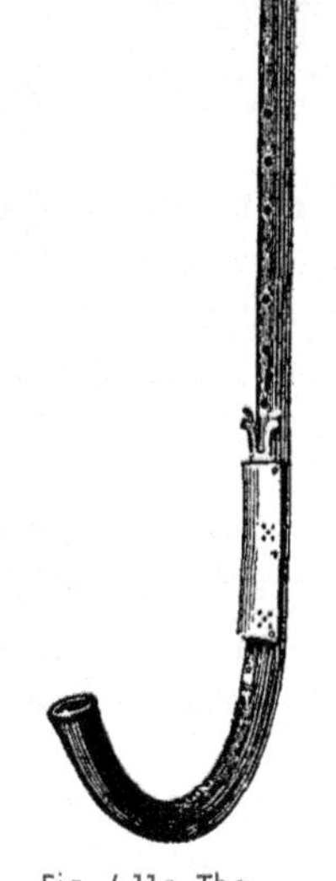

Fig. 4.11c. The capped reed of the crumhorn.

In instruments in which the vibrating reed is the main determiner of pitch, such as a harmonica or reed pipes of an organ, the requirements to the pitch produced by a reed alone are very particular. This reed should demonstrate a very strong preference for one frequency over all others. Such reeds are made of metal, not of cane.

The method of obtaining different notes that is used in almost all woodwind instruments (edgetones and reeds) is the changing of the effective length of a resonator by opening fingerholes or pressing a key, which closes them.

Another way to change the length of resonator is with engaging valves that direct air into additional tubing, increasing the pipe length and lowering the pitch. This method is used in brass instruments. Also in the trombone (which is a brass instrument) a sliding mechanism is used, which lengthens or shortens the vibrating column.

You may also change the frequency of air vibration in the pipe by overblowing it, which means increasing the speed of the airflow blown through the pipe.

We will discuss some of these mechanisms in more detail in Chapters 13 and 14.

4.5 Influence of a Source Size

Our consideration of membranes and strings already brought us to the conclusion that the bigger instrument we have, the lower pitch it creates. Everyday experience confirms this conclusion. A huge tuba definitely creates a lower pitch than a flute piccolo does. Note that size of an instrument is not always simply indirectly proportional with a frequency.

A very important concept in music is an **octave**. If two notes differ in pitches by an octave, they sound very similar to our ear, although one of them has higher pitch. This similarity is emphasized: these two notes have same name, say C, but are of different octaves. For hearing, an octave is an absolutely natural musical **interval**, pleasant for our ear, maybe just a little bit hollow and dull.

So, what is an octave from the point of view of Physics?

Going in pitch one octave higher doubles the frequency.

For instance, the note A^4 corresponds to the vibration of frequency 440 Hz. A^5, one octave higher has, thus, frequency 440 Hz * 2 = 880Hz. The note one octave lower than A^4, A^3, has frequency 440 Hz/2 = 220 Hz.

An octave is subdivided into closer pitches, corresponding to a preferred temperament. In European music we divide octaves into 12 pitches, called semitones. There are also other temperaments, dividing an octave into 15, 17, 19, ... , even 72 pitches. We will discuss in Chapter 7 how to create your own temperament if we are not satisfied with the existing ones.

Example 4.1. What is the frequency of the note two octaves higher than A^3 (220 Hz)?

Answer:

$$440\,Hz \cdot 2 = 880\,Hz$$

The frequency of the note one octave higher is $\frac{440\,Hz}{2} = 220\,Hz$. The frequency of

Fig. 4.12. The saxophone family.

$$220\ Hz \cdot 2 = 440\ Hz$$

the note one octave higher is

$$220\ Hz \cdot 2 \cdot 2 = 440\ Hz \cdot 2 = 880\ Hz$$

For a single family of instruments, for instance, saxophones (Fig. 4.12), which use the same mechanism of sound production and the same form of resonating box or pipe, we find a definite relationship between size and pitch. With good precision for woodwinds, we see that doubling the size of the instrument lowers the pitch by an octave. Note that this kind of comparison **works only** inside a family of **similar** instruments. Comparison of a clarinet and a flute, which have the same size, will bring you to a confusing observation: a flute of the same size sounds one octave higher than a clarinet. These two belong to different families and have different resonators and sound-producing mechanisms. The oboe and clarinet are both reed instruments, but a clarinet of the same length will play approximately half an octave lower. The reason is that the oboe has a conical bore and the clarinet, a cylindrical one. Even small changes in an instrument's construction destroys the simple dependence between its length and the pitch of its voice.

We also see the same in other families of instruments. A cello (string family) plays an octave lower than a viola, but it is less than twice as big. The musical bars (percussion), being cut in half, produce sound that is not one but two octaves higher.

We will consider in more detail the properties of different instruments in Chapters 10–14. So far for us it is enough to understand that **increasing of the size of any instrument will lead to the lowering of its pitch.**

Summary, Terms, and Relations

Steady and transient sounds.

Families of instruments: percussion, strings, pipes.

Change in pitch by one octave corresponds to the doubling of the frequency.

Influence of the size of an instrument on pitch: bigger instruments have lower pitch.

Questions and Exercises

1. What frequency is one octave below 440 Hz? Two octaves below 440 HZ? Two octaves above 440 Hz?

2. How would the pipe organ sound if the temperature in an unheated church is going down?

3. In a certain register of the organ, pipes that produce C_4 (262 Hz) are 60 cm long. What length would you expect for the note C_5? For the note C_3?

4. If you cut a fraction from a cylindrical metallic pipe, what would happen to the natural frequency produced by this pipe?

6. The voice of a tenor saxophone is one octave lower than soprano sax. Which one has a longer body?

5. Using a web search, consider the properties of different keyboards: piano, celesta, and harpsichord. What is the difference between these instruments? What are the common features?

7. Take an empty plastic water bottle and blow air across the narrow opening. Listen to the pitch. Add some water into the bottle and blow air again. What happens to the pitch? How could you explain the raising of a pitch?

8. Take a crystal glass and tap against its side. Listen to its pitch. Now add some water into the glass and tap again. What happens to the pitch? Is there any contradiction with the result of the experiment from the previous question? Discuss the difference and try to explain it.

9. Explain why placing your finger exactly in the middle of a guitar string and pressing it against the neck gives a tone exactly one octave higher than the full-length string.

10. Take a length of flexible plastic tube (garden hose will do) and spin it like a lasso over your head. Change the effective length of your "instrument" by holding it at the end and in the middle. What happens to the frequency of sound created when you spin the tube?

Chapter 5

Measurements of Loudness

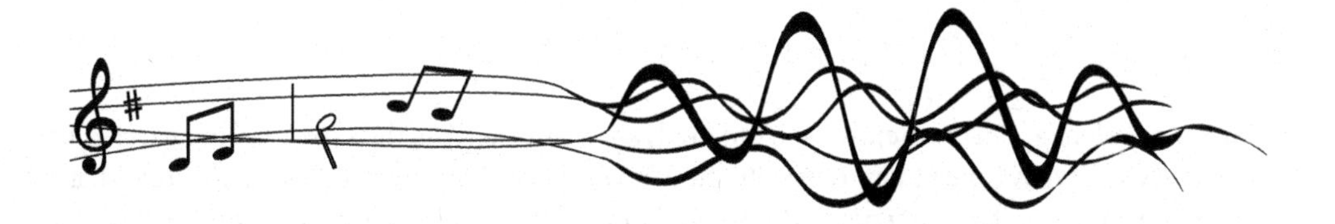

5.1 Power and Intensity

From Chapters 1 and 2 we already know the concepts of amplitude and energy. But how are they connected?

In a sound wave we can consider as the amplitude the maximum displacement of an air molecule vibrating in a wave. Or, what is much more convenient for measurements, we can consider the amplitude of variations of pressure, which can be easily measured. We know also that any wave carries energy. Intuitively it is clear that the bigger the amplitude of pressure variations is, the bigger the energy that is being carried by the wave.

Let us consider Fig. 5.1. There are two pulses shown with amplitudes A and 2A. What will be the ratio of energies in these pulses?

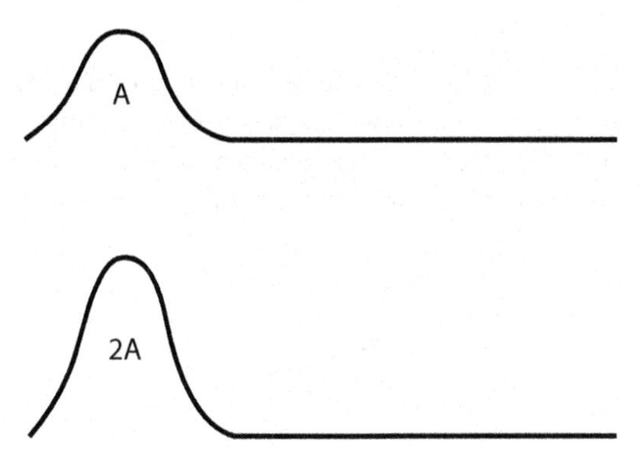

Fig. 5.1. Two pulses of the amplitude A and 2A on the cord.
Source: Sears and Zemansky's University Physics by Hugh D. Young, et al, 2004 Pearson Education, Inc.

Energy carried by the wave is proportional to a square of amplitude.

$$E \sim A^2$$

This means that for a given example the wave (b) will carry not twice, but four times the energy as the wave (a).

Example 5.1. Two waves have amplitudes A and 3A respectively. What is the ratio of energies carried by these two waves? What if the amplitudes were A and 4A?

Answer:

The ratio of energies for waves of amplitude A and 3A is $3^2 = 9$.

The ratio of energies for waves of amplitude A and 4A is $4^2 = 15$.

Another problem appears when we start to analyze what exactly is the energy being carried by the wave? Over what time? In an ideal situation, a wave is an infinite process, so we should at a given point measure the total energy delivered from a source over infinitely long time. And the answer over the infinite time will be infinity. It is not convenient to work with infinities.

So, we introduce a concept that is more useful: power.

Power is the energy point per unit time.

$$P = \frac{\mathcal{E}}{t}$$

The unit of power is the watt (W). You may know this unit from everyday life: all electrical devices are marked with some number of watts, showing how much energy per second this particular device will consume from the electric circuit of your apartment. A typical convectional electric bulb is marked 60–100 W, a typical hairdryer, 1000 W, and so on.

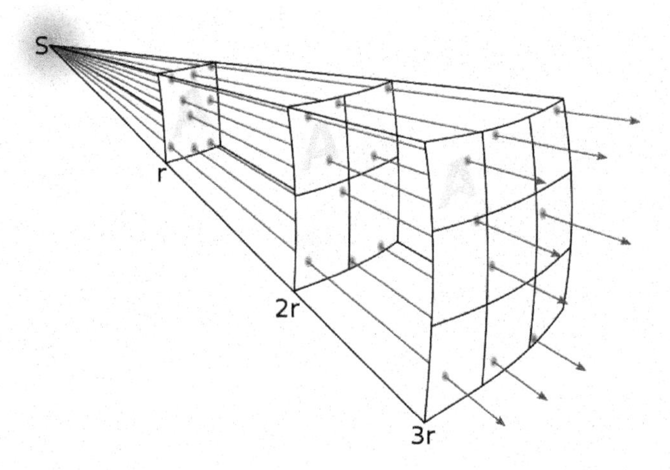

Fig. 5.2. Power delivered in all directions from the point source. Points at the distance 1m, 2m, and 3m are shown as well as the increase of the area element with the distance.

This power is delivered from a source in all available directions, as shown for a source in Fig. 5.2. This source is emitting power P, say 100 W, which is distributed over all the spherical surface shown. But our ear, placed at some point, will catch not all power but only a small fraction of it. Power, as a result, characterizes the ability of our source to emit waves but says little about how much energy is caught by a particular receiver. Now it is time to introduce another useful concept: intensity.

Intensity is the power delivered to a particular point, or

$$I = \frac{P}{S}$$

where S is the area over which our power is distributed. The intensity does not have special units and is measured just in units of power divided by units of area: W/m^2.

Example 5.. Some source of waves emits 100 J of energy during a time of 10 s. What is the power of this source? What is the intensity of this source if it emits energy into an area of 10 m^2?

Answer:

The power of this source is $P = \frac{E}{t} = \frac{100J}{10s} = 10W$. The intensity of this source is $I = \frac{P}{S} = \frac{10W}{10m^2} = 1\frac{W}{m^2}$

Intensity at any given point depends on the amplitude of a wave exactly like the energy considered above: proportional to the square of the amplitude.

5.2 The Inverse-Square Law

We know that increasing a distance from a source obviously decreases the loudness of the sound. This is why at big concert venues amplifiers are used to reinforce sound. Leaving loudness aside for now, let us consider the dependence of intensity on the distance from a source.

A source of small geometrical size, we call it a point source, emits sound uniformly in all directions, as shown in Fig. 5.2. There are two spherical surfaces drawn, one of radius 1 m, another of 2 meters. All points on these spherical surfaces have the same distance from a source equal to the radius of a particular surface. The question is, what is the ratio of intensities at point A on the first surface and B on the second one?

Intensity is indirectly proportional to the area over which power of our source is distributed. In the given example, the area of spherical surfaces can be written as

$$S = 4\pi d^2$$

where d is the radius of the surface, or just distance to a source. So if we increase the distance two times, the area increases four times. From the last formula we see that area is proportional to the distance squared. Thus,

$$I \sim \frac{1}{S} \sim \frac{1}{d^2}$$

So, at point A, shown in Fig. 5.2, the intensity of the sound will be four times higher than the intensity of sound at point B, which is positioned at twice the distance from the source.

The law of dependence of intensity on the distance from a source is called **the inverse-square law**. It describes not only the behavior of sound from a point source with distance but it is also valid for a point source of any nature: the gravitational field outside of a planet, electric field from a charge, intensity of light from a star, and so on.

Example 5.2. The source has a power of 100 W. What is the ratio of intensities at distances 1 m and 4 m from a source?

Answer:

The ratio of distances in our data is 4m/1m = 4. This means that intensity at the distance 1 m is $4^2 = 16$ times larger than the intensity at the distance 4 m.

Example 5.3. Intensity measured 2 m from some sound source is 20 W/m^2. What is the intensity 8 m from this source?

Answer:

The ratio of distances is 8m/2m = 4. It means that intensity at distance 8 m is $4^2 = 16$ times less than the intensity at distance 2 m. As a result, intensity at the distance 8 m is 20 W/m^2/16= 1.25 W/m^2.

5.3 What Is the Problem with Measurement of Sound Levels?

Our ear is an extremely fine and precise device. The ability to perceive sound is really impressive. We can hear a sound with an intensity as small as 10^{-12} W/m^2 (i.e., rustling of leaves in a very quiet forest on a windless day) and easily perceive sound as loud as 1 W/m^2, which corresponds to a jet engine approximately 100 m away. Sure, the sound from a jet engine is too loud, even painful, but we perceive this sensation as a sound.

So, our ear can easily perceive sounds that differ one trillion times in intensity—10^{12} times. But do we really hear such a difference? Of course we say that the sound produced by a jet engine is much louder than the sound of leaves, but not one trillion times.

To bring our measurements closer to human perception, we need another scale, another unit.

5.4 Decibel Scale

For measurements of **sound intensity levels** (SIL, or just sound levels), we use the unit of **decibel (dB)**. This unit is the closest one to actual human sensation, which can be obtained from measurements of pressure variations in a wave. When we say "the closest," we mean that this scale does not take into account the properties of human perception; it is calculated only from the variations of pressure in the sound wave. To get a unit created particularly and only for human perception (unit of loudness), we should first consider the properties of the human ear, which will be discussed more fully in Chapter 8.

First of all, the **decibel** characterizes the ratio between intensities, but not the amount of sound itself. The decibel scale is written through a function, called logarithm (abbreviated "log"), which can found on any calculator, but here we will try to avoid any formulas.

Let us consider two sound sources whose intensities differ 10 times: $\frac{I_2}{I_1} = 10$. We say that the difference between the sound intensity levels (SIL) of these sources is 10 dB: $SIL_1 - SIL_2 = 10$ dB. Note there is not ten times difference in dB. If a third source has intensity 10 times larger than the second, $I_3/I_2 = 10$, then again the difference in the sound intensity levels between the third and second is 10 dB: $SIL_3 - SIL_2 = 10$ dB.

What is the difference in sound intensity levels between the third and first sources considered above? $I_3/I_1 = 100$ and $SIL_3 - SIL_1 = 20$ dB. So, if the ratio between intensities of two sources is just 1 with n zeros after, the difference in sound intensity levels is n*10:

$$\frac{I_2}{I_1} = 10; \; SIL_2 - SIL_1 = 10dB$$

Example 5.4. Two sources of sound have intensities 1 W/m² and 0.001 W/m² and a distance of 1 m. What is the difference in sound intensity levels of these two sources?

Answer:

The ratio of intensities of these two sounds is $\frac{I_2}{I_1} = \frac{1 W/m^2}{0.001 W/m^2} = 1000$. This means that the difference between sound intensity levels is $SIL_2 - SIL_1 = 30dB$.

If the ratio of intensities is not a simple power of 10, we consult Table 5.1

Table 5.1. Intensities ratios and differences in SIL.

Intensity ratio I_1/I_2	*Sound Intensity Level Difference SIL_1-SIL_2 (dB)*
1	0
1.6	2
2	3
4	6
5	7
8	9
10	10
100	20
1000	30
10^n	$10n$

Example 5.5. Two sources produce sounds with an intensity ratio of 20. What is the sound level difference of these two?

Answer:

The intensity ratio can be written as 20 = 2 x 10. Now, check Table 5.1. Ratio 2 gives us the sound intensity level difference of 3 dB, and the ratio 10 corresponds to SIL difference of 10 dB. The result:

$$SIL_2 - SIL_1 = 10dB + 3dB = 13dB$$

Example 5.6. The sound intensity level difference of two sources is 26 dB. What is the ratio of their intensities?

Answer:

Table 5.2. Typical SILs with corresponding intensities.

Sound Source	SIL (dB)	I (W/m²)	Human reaction
Jet engine at 10 m	150	10^3	unbearable
Rock Concert	110	0.1	Musically useful range
Subway train	90	0.001	
City Traffic	70	10^{-5}	
Quiet Conversation	60	10^{-6}	
Library	40	10^{-8}	
Whisper at 1 m	20	10^{-10}	

Let's write the difference between SILs in the form $26dB = 20dB + 6dB$. Now check Table 5.1. The SIL difference 6 dB corresponds to the intensity ratio 4 times, and the SIL 20 dB corresponds to 100 times the ratio of intensities.

The result:
$$\frac{I_2}{I_1} = 4 \cdot 100 = 400\,times$$

From these two examples we see the important rule that we always must follow while working with intensities and SILs:

If intensities are multiplied, SILs are added. Never multiply sound intensity levels!

When we read the manual for a loudspeaker, we see something like "A distance of 1 m provides an intensity level of 90 dB." What does this mean? In this phrase 90 dB characterizes a ratio between the actual intensity of this loudspeaker and some standard, $I_0 = 0.000000000001 = 10^{-12}$ W/m^2. I_0 is an intensity of an extremely soft sound—which can be detected under ideal conditions by a very good human ear—called the "threshold of human hearing." So, 90 dB in a manual of a loudspeaker means that the intensity of this loudspeaker measured at distance of 1 m is 10^9 higher than the standard I_0.

The sound levels typical for some environments are shown in Table 5.2.

Example 5.7. One car engine in a garage creates SIL of 80 dB. What will the SIL created by 10 such engines be?

Answer:

To answer this question we should:

1. Keep in mind that SILs **should not be multiplied**. So, the answer cannot be 80 * 10 = 800 dB; this is well beyond any bearable sound.

2. We **should multiply** intensities:

   If the intensity of one engine is I, then the intensity of 10 identical engines is 10 x I, so the ratio of intensities is 10.

3. If the ratio of intensities of 10 engines and one engine is 10, then the difference between the sound intensity levels is 10 dB. That's it: 10 engines will produce sound louder than 1 engine by 10 dB.

4. The answer is $SIL_{10} = 80dB + 10dB = 90dB$.

As we see from Table 5.2, sound levels important for music are between 30 and 100 dB. But it is worth mentioning that levels below 50 dB are seldom used because of ever-present background noise. Usually music below 40 dB is barely audible.

5.5 Sound Intensity Level and Distance from Source

We see from subchapter 5.3 that intensity of sound decreases indirectly proportional to square distance.

If we measure the intensity at distance $d_1 = 2$ m and d_2=4 m, distances increase twice, but the ratio of intensities is $I_1/I_2 = 4$. What is the change in the sound intensity level? We consult with Table 5.1 and see that decreasing intensity 4 times corresponds to the difference in sound intensity levels for 6 dB. If we compare SIL at 2 m with the SIL 8 m, the ratio of intensities is 16 = 10 * 1.6, so the difference in SIL is 10 + 2 = 12 dB.

Example 5.8. At a distance of 1 m, the sound intensity level is 100 dB. What is the sound intensity level at a distance of 100 m from a given source?

Answer:

The ratio of distances is 100 m/1m = 100. This means that the ratio of intensities at these points is $100^2 = 10000$. So, the SIL at a distance of 100 m should be 4 x 10 = 40 dB less than at distance 1 m.

$$SIL_{100m} = SIL_{1m} - 40dB = 100dB - 40dB = 60dB$$

Example 5.9. At a distance of 10 m, a jet engine produces a sound of 130 dB. At what distance will SIL be equal to the somewhat bearable 110 dB?

Answer:

The SIL difference is 130 dB - 110dB = 20 dB. It means that intensity should decrease (check Table 5.1) $10^2 = 100$ times. Intensity is indirectly proportional to the distance square. This means that when intensity is decreased 100 times, the distance should increase $\sqrt{100} = 10$ times.

As a result, the distance at which the SIL is 110 dB is 10m x 10 = 100 m.

5.6 Noise Pollution

Environmental noise, the sum of noise from transportation, construction, and other activities around us can damage human health. Noise pollution can cause aggression, annoyance, headaches. Chronic exposure to noise may cause noise-induced hearing loss. The U.S. Department of Labor's Occupational Safety and Health Administration (OSHA) has set many standards for noise in offices, libraries, factories, and other facilities.

In dwellings, the critical effects of noise are on sleep, annoyance, and speech interference. The standard for bedrooms is 30 dB for continuous noise and 45 dB for single events. To protect people from being seriously annoyed during the daytime, the SIL of outdoor living areas cannot exceed 50 dB.

For schools, the critical effects of noise are on speech interference, disturbance of provided information, communication, and, again, annoyance. The background noise in classrooms should not exceed 35 dB during teaching hours.

During festive events it is impossible to avoid high-level noise, usually caused by fireworks, impulsive sound, and music from all sides. Music itself is good, and it could be pretty loud, but several pieces of high intensity music mixed together are as good as just noise. The sound level of rock concerts is typically in excess of 100 dB. And frequent attendance at such noise exposure could lead to serious hearing problems.

Another concern is the health of employees at loud events. They should not be exposed to sound levels greater than 100 dB during a 4-h period for at most four times per year. To avoid hearing impairment, the maximum sound level should never exceed 110 dB.

Summary, Terms, and Relations

Intensity of the source is the power delivered to a particular point, or

$$I = \frac{P}{S}$$

Intensity of the wave is proportional to a square of amplitude.

$$I \sim A^2$$

The law of dependence of intensity on the distance from a source is called **the inverse-square law:**

$$I \sim \frac{1}{S} \sim \frac{1}{d^2}$$

For measurements of **sound intensity levels** (SIL, or just sound levels) we use the unit of **decibel (dB)**. Use Tables 5.1 and 5.2 for answering questions.

Important: If intensities are multiplied, SILs are added. Never multiply sound intensity levels!

Questions and Exercises

1. The sound intensity of $I = 10^{-6}$ W/m^2 falls on an area of 1 cm^2. What is the total power P in watts delivered to this area? If the sound continues for 1 min, what total amount of energy in joules is delivered?

2. What is the intensity I of the sound with intensity level 60 dB? 90 dB?
3. What SIL in dB must you have to get a sound intensity 1 W?
4. If two sources produce sound levels of 55 dB and 65 dB, what is the ratio of intensities of these sources at a given point?
6. If the violin produces a reading of 70 dB on your sound level meter, what reading will be produced by two identical violins? By ten violins? By 20 of them?
5. The intensity of a sound 10 m from its source is 10^{-5} W/m^2. What is the intensity at a distance of 100 m from the source?
7. What are the SILs at the distances 10 m and 100 m in the previous example?
8. At the distance 5 m from a sound source, the SIL is 80 dB. What will the SIL be at a distance of 10 m from the source?
9. Consider two sound sources of 60 dB and 80 dB. What is the combined sound level (SIL) of these two sources?
10. At a distance 10 m, the engine of a jet produces sounds of 150 dB, which is unbearable. At what distance might you tolerate 110 dB?

CHAPTER 6

Physical Properties of Sound

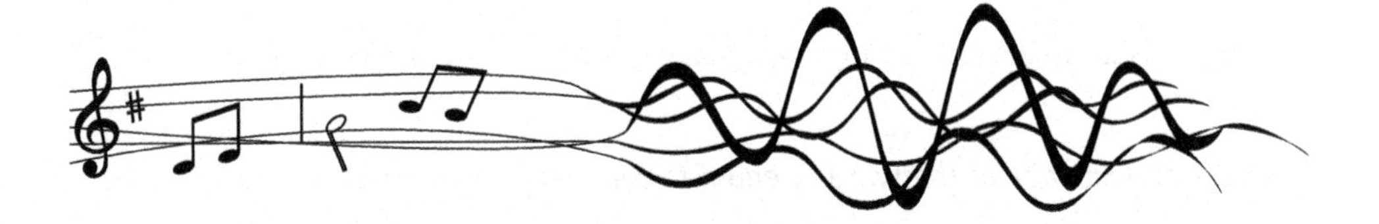

6.1 General Remarks About Wave Propagation

The properties of waves that we will consider in this chapter **are not** features of sound waves alone. All laws that we will discuss are applicable to any type of waves, for example, electromagnetic (light, radiowaves, heat, microwaves, etc.).

Another important thing to understand: for it to be possible for us to see light or to hear sound, these waves must hit the organs which are responsible for our particular sensations. We will not see the light, maybe even very powerful light, created by a laser unless this light hits our eye. We will not hear any sound if the sound wave does not hit our eardrum.

It is worth repeating, although we have considered this in Chapter 3, that for sound to travel, it **needs a medium**. Only electromagnetic waves can travel through a vacuum. Because we usually need air for purposes of music, in this particular chapter we will consider sound waves propagating in the air unless something else is stated.

6.2 Law of Reflection

Sound waves practically never travel in an infinite medium. Consider a lecture hall—there is no infinity at all: all the sound is bouncing back from the walls, ceiling, floor, like some kind of hard ball. We are saying: **sound is reflected**. If you stay 50–100 m from a hard wall on a quiet street and scream, you can hear the sound from your scream echoing from the wall. If the wall is hard enough and there are no collateral disturbances in the air, the echo is absolutely clear, and the sound seems to come from the area behind the wall, the way your image in a bathroom mirror seems to stand behind the mirror. Note that the distance from you to the wall should be pretty substantial to hear the echo: if the distance between you and the wall is small, the echo surely also exists, but our ear does not resolve these two close signals: original and reflected. They seem to us to be one pulse of sound.

Sound reflects from any surface or any obstacle. If the surface is smooth and hard, as, for instance, walls in a marble lobby, the reflection will be regular or **specular** (Fig. 6.1a). All the sound, like a beam of light from a laser, is reflected from this surface without losing its shape.

And here we come to the first law for propagating waves—the law of reflection:

The angle of incidence of the wave is equal to the angle of reflection.

If a wave falls on the smooth surface perpendicular to it, it will be reflected straight back. If there is some other angle between the surface and the direction of travel of our wave, it reflects back, creating the same angle with a surface as the angle of incidence.

But the situation of specular reflection is pretty rare for sound waves. For light, this kind of reflection occurs when we are looking into a plain mirror. Much more often, however, we deal with surfaces that are not smooth. They have bumps and imperfections. As a result, the wave is reflected from such a surface in all directions, as shown in Fig. 6.1b. Note that the law of reflection holds here for each element of a considered wave, but all angles of incidence are now different as well as all angles of reflection.

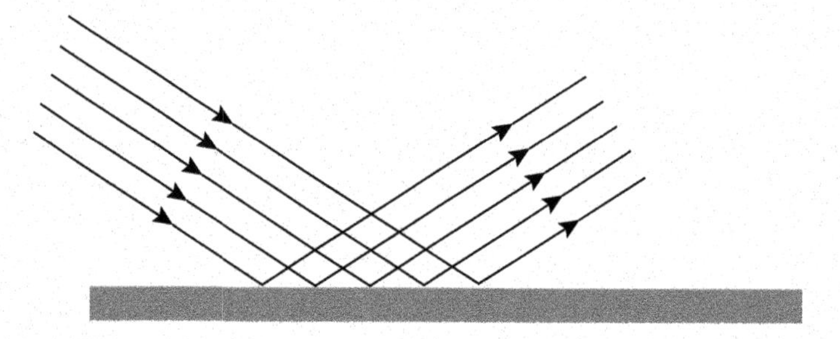

Fig. 6.1a. Reflection from the smooth surface.

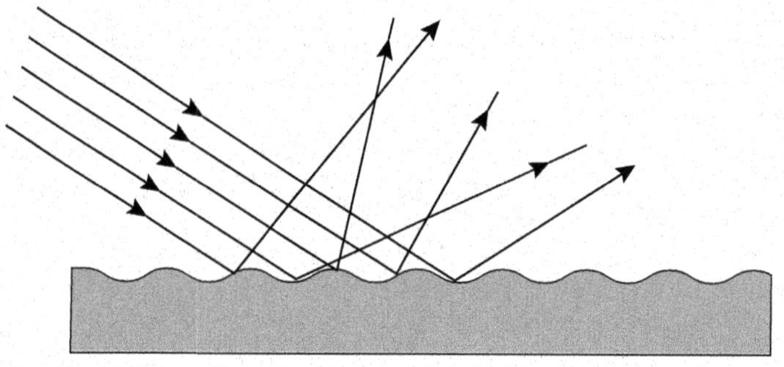

Fig. 6.1b. Reflection from the rough surface with significant imperfections.

When we use everyday words such as "smooth" or "bumpy" in Physics, we should always understand with what parameter of the wave we should compare the average size of the bumps. **The wall is smooth for a wave if the average size of the imperfection (bump) is much less than the wavelength of the wave.** For instance, all surfaces that could serve as mirrors have characteristic imperfections less than a thousandth of a millimeter because the wavelength of visible light is very short. As a result, all mirrors feel polished to us; we cannot even feel any imperfection with our fingers. For sound this rule is also true. But the wavelength of an audible sound is between 2 cm and 20 m, so bumps can be much more than for light, and a surface will still provide specular reflection. It requires irregularities of several centimeters in size to scatter treble waves and much larger irregularities to scatter bass notes.

Multiple reflections from different surfaces allow sound to travel from one room into another (Fig. 6.2); this is why we can often hear people talking in another room even without seeing them.

Even a very hard wall does not reflect all the original sound. They say no one real surface is a **perfect reflector**. Sound, being an elastic wave, sets any surface into very tiny oscillations. Even marble, which is very hard, has some "give." This reaction of a surface drains energy from the original sound signal; it **absorbs** part of original sound. In principle, waves inside of material can travel to another boundary. We witness this when we hear through the not-very-thick wall.

Fig. 6.2. Multiple reflections allow a listener to hear the sound in an adjacent room.

For the study of acoustic phenomena, scientists are using so-called anechoic chambers. These are rooms with perfectly absorbing walls. The sensations in such a room are pretty weird: you can hear your friend, for instance, only when he or she is almost facing you; all sounds are muted because there are no reflections from walls. You feel like you suddenly became partially deaf.

6.3 Law of Refraction

We all know that a pencil in a glass of water looks strangely broken (Fig. 6.3). This occurs because the speed of light in the air and water is different: light moves in water approximately 30% slower. Another phenomenon, for instance, which everybody can observe while driving along a straight highway on a hot day, is that the car far away from you appears to move

Fig. 6.3. Refraction of the light: the pencil looks broken.

on a puddle of water and is reflected in this puddle upside down. If the distance between cars is decreasing, the "puddle" evaporates quickly, and you see just a car moving on absolutely dry pavement (Fig. 6.4). We say that this is highway mirage.

Fig. 6.4. The highway mirage during the hot summer day.

This phenomenon, called **refraction**, occurs always when the speed of the wave (in these examples, of light) depends on position. Because of this dependence, light in the example with the car did not move along a straight line but followed a curved path.

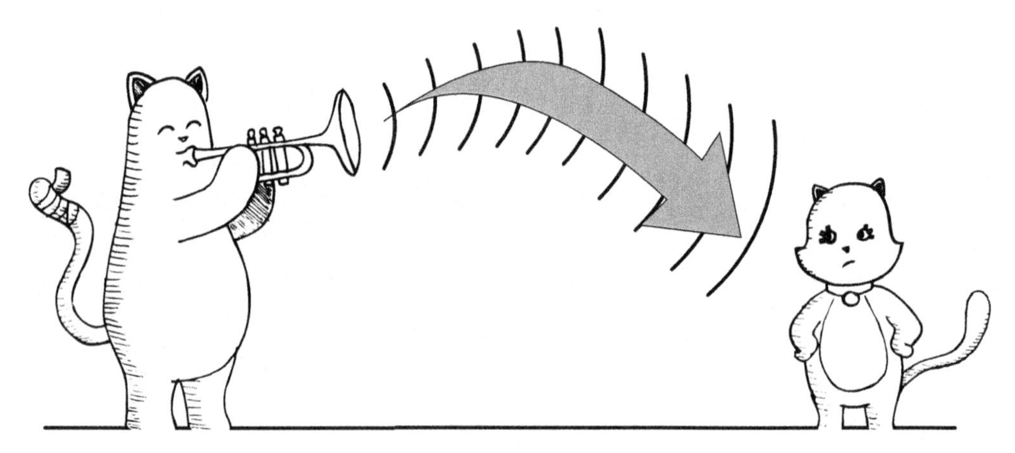

Fig. 6.5. The sound bends down when the temperature of the air increases with the altitude.

We do not consider sound traveling from air into the water because it is not musically useful. But the speed of sound strongly depends on temperature. Here we can expect that a change of temperature with distance from the ground can cause the bending of sound waves up or down.

If the temperature near the ground is increasing with the altitude, so does the speed of sound. Such conditions may occur during a winter day. The sound wave in such conditions behaves as shown in Fig. 6.5, bending toward the ground. The sound travels slower at the ground, and as a result, the upper levels "outrace" the lowest ones. Such a situation, however, is pretty rare.

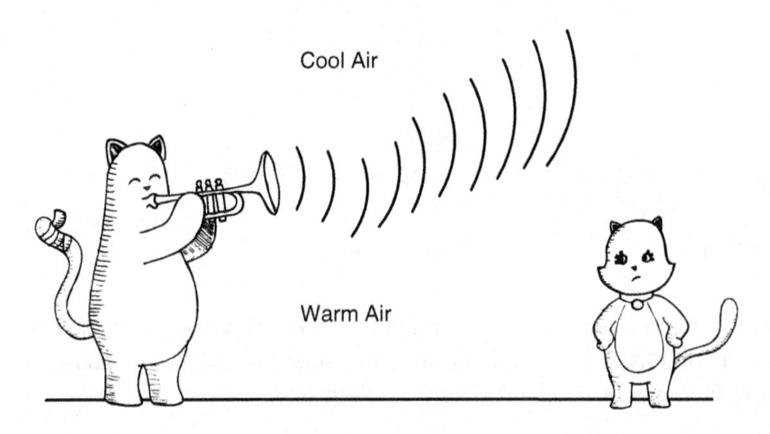

Fig. 6.6. The sound bends up when the temperature of the air is decreasing with the altitude.

In more common conditions, the temperature of air gradually decreases with the altitude and so does the speed of sound. Then the sound near the ground travels faster (Fig. 6.6), and the sound wave bends up, away from the ground. The listener, shown in the picture, will receive just a little sound or none.

Wind also influences the behavior of sound waves. The speed of wind usually increases with altitude, and near the ground the air is almost still. As a result, wind carries along the higher parts of a wave, together with air, in which our sound wave is traveling. If that same listener in Fig. 6.7 stands downwind from a source, the wind brings sound toward him or her more quickly, bending the wave toward the ground. Meanwhile, a person standing upwind hears very little from a distant source as the sound wave for him or her is bent upward.

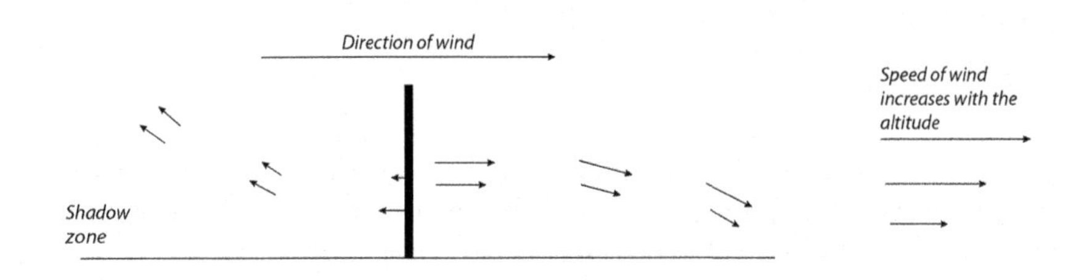

Fig. 6.7. The sound bends down when "downwind" from the source and bends up when "upwind" from the source.

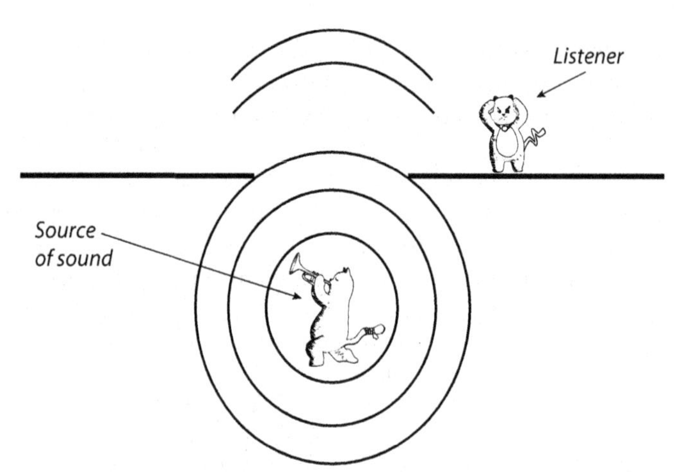

Fig. 6.8. Example of diffraction: this listener can hear sound from a source hidden behind the wall.

6.4 Diffraction

We all know that we can hear sound from around corners. We already discussed one reason for this: multiple reflections. But even with no reflecting surfaces in sight, we can hear sound when the source of which is hidden behind a finite wall (Fig. 6.8).

The spreading of waves behind the obstacle, wall, or behind the opening (Fig. 6.9a) is called **diffraction** and, as was mentioned above, is the general property of all kinds of waves. You can easily observe diffraction of water waves spreading behind some obstacle, such as a rock.

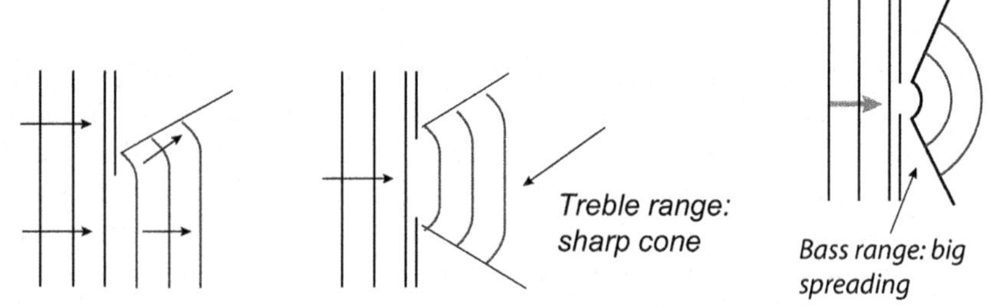

Fig. 6.9a. Diffraction: spreading of sound waves behind the obstacle

Figs. 6.9b and c. Spreading of waves strongly depends on frequency. The treble range waves (b) spread less than the bass range waves (c).

How much the wave diffracts depends on the comparison between a wavelength and the characteristic size of an opening or obstacle. If the wavelength is smaller than the size of an opening (Fig. 6.9b), this wave will spread into a pretty sharp cone. Such a situation occurs for treble sound going through a door. If the wavelength is pretty big, the same size as the opening, or even bigger, the spreading behind the opening is almost uniform in all directions (Fig. 6.9c). Thus, again, as in the situation with reflection, we come to a comparison of two lengths: the

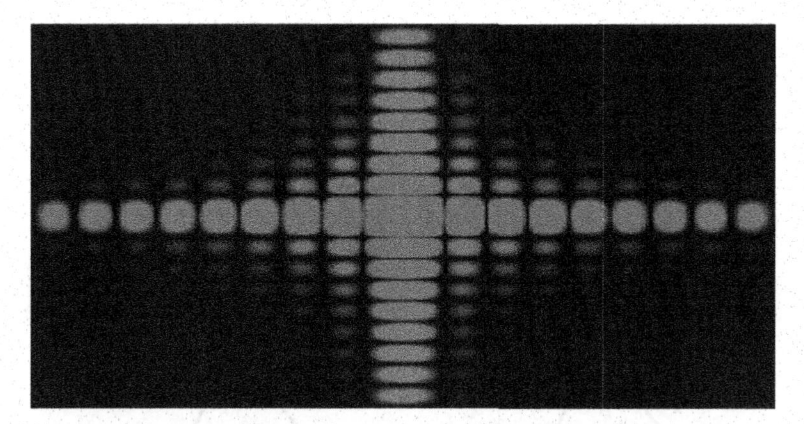

Fig. 6.10. Diffraction of light on a small obstacle. Note the numerous maxima and minima of illumination.

wavelength, which is a characteristic of a wave, and the size of an obstacle (opening or bump for reflection), which is a characteristic of a surface with which our wave interacts.

In everyday life we never observe the diffraction of light. The reason was just discussed above: for diffraction to be significant, we would need to have an obstacle of the size of the light wavelength, which is very tiny. For instance, we could observe diffraction on a single human hair. But even for this experiment you cannot use just sunlight, you need to use a laser (Fig. 6.10).

Diffraction is the reason why sound, coming from a mouth of a person, spreads in different directions: so you can hear this person even when you are not exactly facing him or her. Sound spreads from the sound hole of a guitar, from the f-holes of a violin. But a violin usually has a pretty high voice (short wavelengths), so the spreading of the sound of a violin goes into a sharp angle originating on a top plate of an instrument. Thus, the violin should be turned toward the audience with its top plate. This is the reason why, in big orchestras, violinists always sit to the left of a conductor. Meanwhile, cellos have a lower voice, spreading around the instrument more uniformly.

Because of diffraction, we hear with our left ear the sound that is coming from the right. Bass notes have wavelengths much bigger than your head, so their strength is practically the same for both ears. Meanwhile, treble notes can be much weaker at the left ear than at the right.

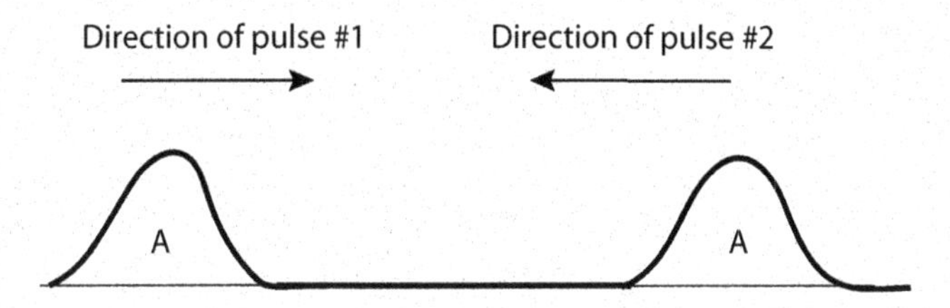

Fig. 6.11a. Two identical pulses move towards each other.

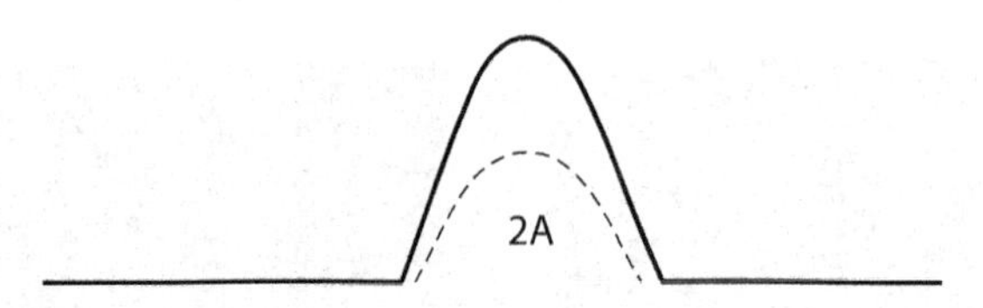

Fig. 6.11b. The amplitude of the wave at the very moment of the meeting of two pulses is the sum of amplitudes.

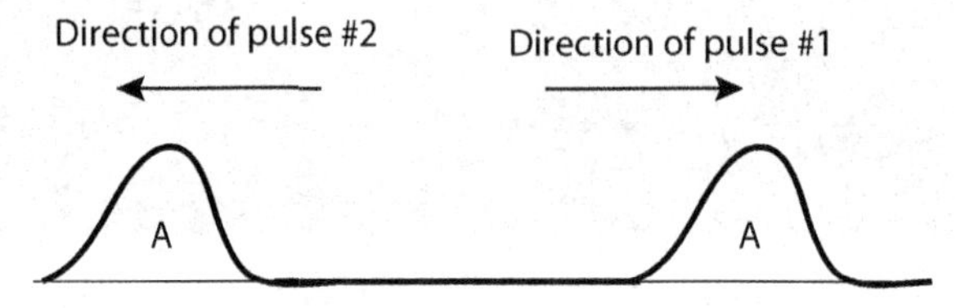

Fig. 6.11c. Pulses continue their way unchanged.

6.5 Interference

So far we have considered situations with only one wave. It can be reflected or refracted, but it was alone. But what will happen if two waves encounter each other while propagating through a medium?

For starters, let us discuss the simplest situation: two pulse waves on an elastic cord (rope, string), moving toward each other, as shown in Fig. 6.11a. We created two identical waves (1) and (2),

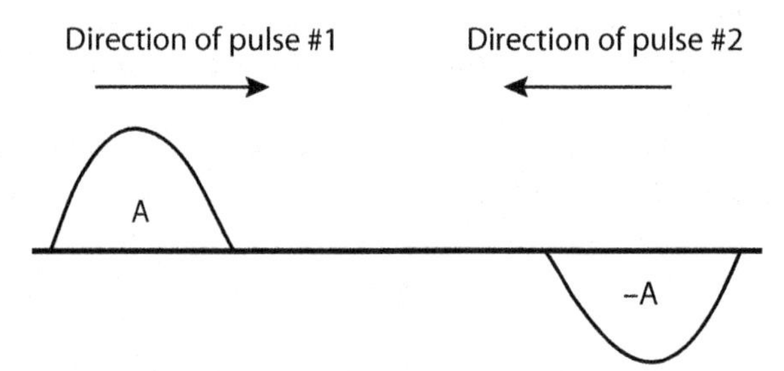

Fig. 6.12a. Two pulses of identical amplitude but created in opposite directions move towards each other.

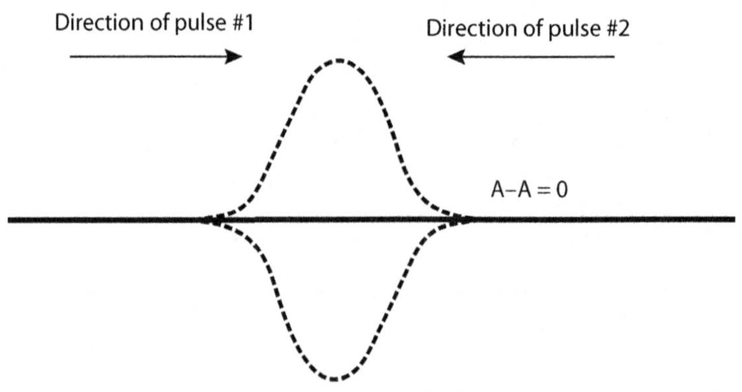

Fig. 6.12b. At the point of meeting, these pulses completely cancel each other.

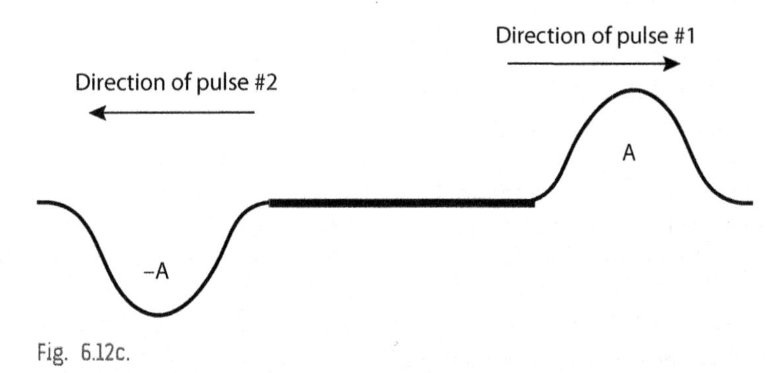

Fig. 6.12c.

(it is not important, but simpler for consideration) both with amplitude A. In the middle of a cord these two meet each other. The amplitude of the resulting wave at the very moment of this meeting is A + A = 2A, twice as big as each of them (Fig. 6.11b). But waves do not stop at this point; they continue their motion, absolutely unaffected (Fig. 6.11c).

This is an example of **interference**. In the above-mentioned situation, we considered the **constructive interference** because we had pulses in the same direction, and the result is bigger than the amplitude of each of these waves alone.

Now we create pulses on the rope in opposite directions (Fig. 6.12a). When these two meet in the middle of a cord, the pulse for a moment disappears because the amplitude of a resulting wave is A – A = 0 (Fig. 6.12b). But a moment later, waves appear unaffected again, continuing their motion (Fig. 6.12c). This is an example of **destructive interference**. Note: waves do not disappear into nothing and appear from nothing. Any wave is carrying energy, which is nearly conserved. When waves cancel each other, all the energy of this process is stored in the elastic tension of a cord.

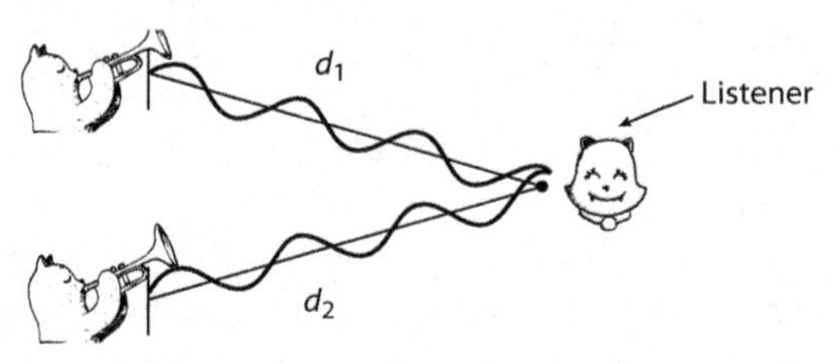

Fig. 6.13a. Constructive interference: the path traveled by the wave from the first source is exactly equal to the path covered by the wave from another source.

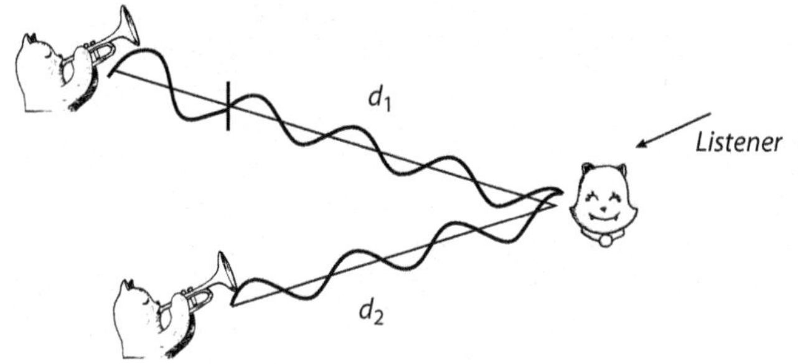

Fig. 6.13b. One of the sources moved exactly one wavelength back. The difference in paths traveled by waves from the first and second source is now one wavelength. Waves still experience the constructive interference at the listener's ear.

Now we apply this knowledge to sound waves. We arrange two loudspeakers, as shown in Fig. 6.13a. Let the loudspeakers produce a signal of frequency f and wavelength λ. This is the same signal that was just sent to two loudspeakers, which means the crests and troughs of a sound wave appear from the loudspeakers simultaneously, **in phase**. A listener is standing exactly at the midline between the speakers. Distance between an ear and each loudspeaker is the same, d. So, the wave from each loudspeaker travels the same distance to the ear of our listener, arriving also in phase. This listener hears an enhanced signal. Here is a point of constructive interference.

But are the points of constructive interference all positioned only at the midline? To answer this question, let's move one of the loudspeakers back for a distance exactly equal to the wavelength of our signal (Fig. 6.13b). A sound wave is infinite and periodic, so a listener again hears the enhanced signal. But the distance traveled by two waves differs now by one wavelength. We can move the loudspeaker back another wavelength, and again get constructive interference; the distances traveled by the two waves differ now by 2 wavelengths, and so on.

Thus, condition of **constructive interference** should be written in such a form:

$$d_1 - d_2 = m\lambda, \text{ where } m = 0, 1, 2, 3\text{—any integer.}$$

Example 6.1. What are the wavelengths and frequencies of sound waves that demonstrate constructive interference at a point 4 m from one loudspeaker and 6 m from another?

Answer:

The condition of the constructive interference shown above says that the first wavelength should satisfy the condition:

$$d_1 - d_2 = 1 \bullet \lambda_1, \text{ or with our data:}$$

$6.0m - 4.0m = \lambda_1$, which gives us the answer 2.0 m. The next wavelength can be obtained from the condition

$d_1 - d_2 = 2 \bullet \lambda_2$, which gives the wavelength of 1 m. The frequencies should be obtained from the relation between the speed of sound, wavelength, and frequency:

$v = \lambda f$, so $f = \frac{v}{\lambda}$. Let's take the speed of sound at temperature 20°C, then

$$f_1 = \frac{v}{\lambda_1} = \frac{344 m/s}{2m} = 172 Hz$$ The second frequency is

$$f_2 = \frac{v}{\lambda_2} = \frac{344 m/s}{1m} = 344 Hz$$

To consider destructive interference, we will arrange loudspeakers as shown in Fig. 6.14a. Now the wave from loudspeaker 1 should travel to the ear of a listener at a distance larger than the distance for the wave from loudspeaker 2 exactly by **half of the wavelength**. As a result, the ear trough from one wave meets the crest of another. The waves are **out of phase**, and they cancel each other. We can move the first loudspeaker one wavelength back, as in the situation

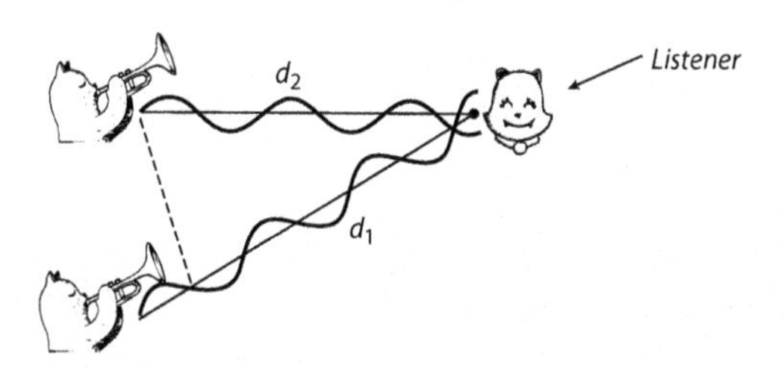

Fig. 6.14a. Destructive interference: the path traveled by wave from first source is half of wavelength bigger than the path travelled by the wave from another source.

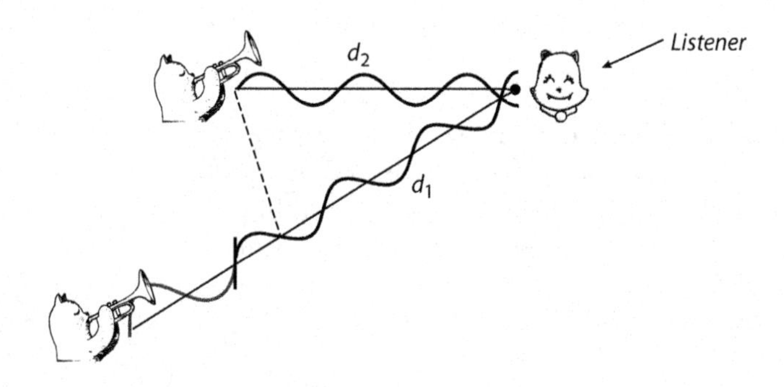

Fig. 6.14b. One of the sources moved exactly one wavelength back. Waves still experience the constructive interference at the listener's ear.

considered above (Fig. 6.14b) and again get a destructive interference. The general condition of **destructive interference** should be written as:

$$d_1 - d_2 = m\lambda + \lambda/2, \text{ where } m = 0, 1, 2, 3\text{—any integer.}$$

Example 6.2. What are the wavelengths and frequencies of sound waves that demonstrate destructive interference at a point 4 m from one loudspeaker and 6 m from another?

Answer:

The condition of the destructive interference says that the first wavelength should satisfy the condition:

$$d_1 - d_2 = \lambda/2, \text{ or with our data:}$$

5.0 m – 4.0 m = $\lambda_1/2$, which gives us the answer 4.0 m. The next wavelength can be obtained from the condition

$d_1 - d_2 = (\lambda_2 + \lambda_2)/2$, which gives the wavelength of 1.33 m. The frequencies should be obtained from the relation between the speed of sound, wavelength, and frequency:

$v = \lambda f$, so $f = v/\lambda$. Let's take the speed of sound at temperature 20°C, then

f_1 = 344 m/s/4 m = 86 Hz. The second frequency is

$$f_2 = 344\ \text{m/s}/1.33\ \text{m} = 258\ \text{Hz}.$$

To conclude, we have numerous points of destructive and constructive interference from two loudspeakers, as can be seen in Fig. 6.15.

We considered the simplest situation demonstrating the interference of waves. Note that interference occurs not only for waves of the same frequency but for all sound waves which encounter each other in space.

Observation of the interference of sound waves in everyday life is not an easy task. The best way to arrange such an experiment is to set it up in an anechoic room, as discussed earlier. In real life sound reflects from all surfaces around us, creating a pattern that is not clear and maybe even with a completely "averaged" interference pattern.

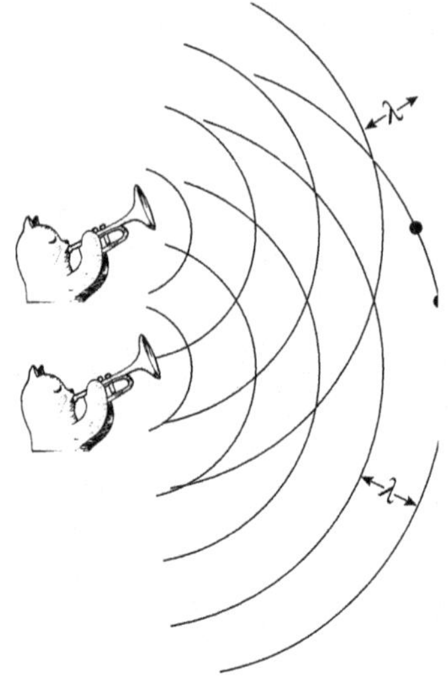

Fig. 6.15. Two sound sources. The waves intersect at numerous locations creating many points of constructive and destructive interference.

6.6 Beats

We always observe interference if two waves from different sources come to a given point. In Subchapter 6.5 we discussed a particular situation when the amplitudes and frequencies of two waves were the same.

But what happens if waves do not have exactly the same frequency? The result is shown in Fig. 6.16. We are adding two waves of slightly different frequencies, f_1 and f_2, and are again getting a

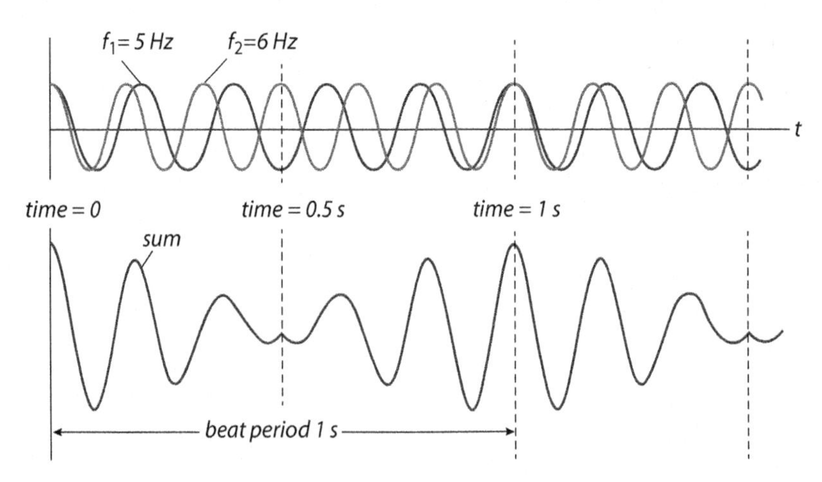

Fig. 6.16. Creation of beats: two waves with frequencies 5 Hz and 6 Hz (upper graph) have been added to create the wave shown in the lower graph, demonstrating the periodic oscillations of loudness.

periodic wave pulsating with a frequency $f_1 - f_2$. This kind of interference has the special name, **beats**, because for two sound waves close in frequency, you can clearly hear these pulsations of loudness. You hear not two sounds, like a chord on a piano, but one whose intensity rises and falls repeatedly. Note that it is impossible to say which frequency is higher.

If we mix not two, but three or more waves, the beats occur at all possible differences between frequencies.

> Example 6.3. What beat frequency do we hear from two tuning forks at 440 and 444 Hz? From three tuning forks at 440, 442, 445 Hz?
>
> Answer:
>
> To find the frequency beats, we should take all possible differences between given frequencies, which gives us 2 Hz, 5 Hz, and 3 Hz.
>
> Example 6.4. Two tuning forks demonstrate frequency beats of 2 Hz. One of them is marked 440 Hz. What is the frequency of the other one?
>
> Answer:
>
> We cannot say which frequency is higher; the human ear cannot resolve frequencies so close to each other. So, the answer is that the second tuning fork has frequency 442 Hz or 438 Hz.

Beats always occur when we mix sounds of two or more frequencies. Then why we do not hear beats while hitting a chord in, say, the midrange of a piano? Because the difference in frequencies of adjacent keys of a midrange is pretty big—more than 20 Hz. Our ear simply does not resolve

Fig. 6.17. Police car at rest: both observers hear the same frequency of a siren actually produced by the car.
Adapted from Sears and Zemansky's University Physics by Hugh D. Young, et al, 2004 Pearson Education, Inc.

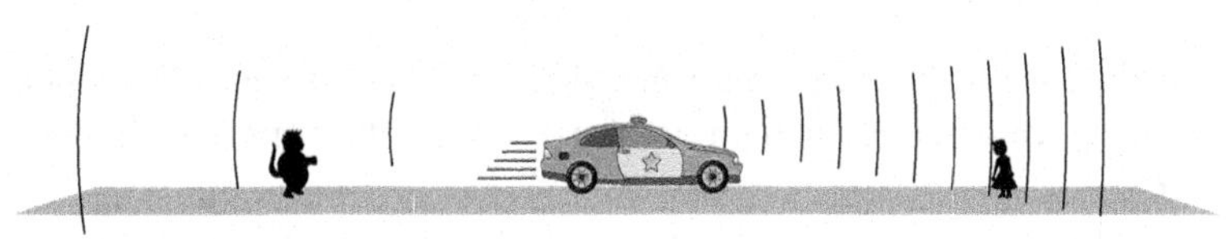

Fig. 6.18. Police car moves: the observer in front hears the apparently higher pitch, the observer behind the car hears the apparently lower pitch.
Adapted from Sears and Zemansky's University Physics by Hugh D. Young, et al, 2004 Pearson Education, Inc.

such rapid oscillations; the beats cannot be heard individually, giving us a sensation of two different sounds played together. But beats can be clearly heard while playing the lowest organ register.

Beats are used in tuning musical instruments. The more rapid beats you hear between a piano string and a tuning fork, the greater the difference between frequencies. So, the musician is changing the tension of a string, making the frequency of beats smaller and smaller, bringing the pitch of a string closer and closer to a pitch of a tuning fork. It is practically impossible to hear beats slower than 3 s in period (1/3 of Hz). One half Hz passes as a good tuning. As we will see in Chapter 7, the human ear does not hear if a frequency is 0.5 Hz off perfect tuning.

Example 6.5. Two strings demonstrate a beat frequency 2 Hz. What would happen to the frequency of beats if we decrease tension in a string with lower pitch?

Answer:

When we decrease the tension of a string demonstrating lower pitch, the pitch of this particular string is getting even lower. As a result, the frequency difference between these two strings increases and the frequency of beats increases too.

6.7 Doppler Effect

If a source producing sound and an observer receiving it are at rest relative to each other, the sound wave reaching an observer has the same frequency as the frequency of a vibrating element of a source. In Fig. 6.17 two observers are at rest with a police car, so they hear the same pitch of a siren produced by the car.

If this police car moves (Fig. 6.18), the sound waves around it are changing their shape: in front of the police car, the distances between crests become smaller, and, as a result, the wavelengths are smaller too. Behind the car wavelengths of the produced sound become bigger. As a result, an observer in front of the car meets crests more frequently and hears a pitch higher than the actual one. The observer behind the car hears a lower pitch.

This apparent shift in frequency is called the **Doppler effect.**

The Doppler effect always occurs when the distance between the source and observer changes, no matter what the reason is for this change: motion of a source, motion of an observer, or both. If distance **decreases**, an observer hears **higher** pitch, if distance **increases**, an observer perceives **lower** pitch.

Note that when distance is changing uniformly—if, for instance, a source is moving with constant velocity—the pitch is neither higher nor lower but **stays the same**. It does not gradually go higher or lower.

If the relative speed of the source and observer V is much slower than the speed of a signal v (Note: this true for everyday speeds when we are talking about sound and always true for speeds of light), the frequency shift could be written as

$$\left|f_{received} - f_0\right| = \frac{f_0}{v} V$$

where f_0 is the frequency actually emitted by a source.

Example 6.6. You have two tuning forks, each marked 440 Hz. Leaving one of them at rest, you start to move away at a speed of 2 m/s, keeping the second tuning fork in your hand. The beats of what frequency do you hear?

Answer:

When you move away from a tuning fork at rest, the apparent pitch of that fork sounds lower for you by

$$\frac{f_0}{V}v = \frac{440\,Hz}{344\,m/s}2\,m/_{s} = 2.5\,Hz$$

The beats between the tuning fork in your hand and the tuning fork at rest are 2.5 Hz.

The Doppler effect, as well as all other phenomena considered in this chapter, are the properties of many kind of waves. Astronomers use the described effect to study the motion of distant stars and to define the speed of this motion by Doppler shift in their color. The low frequencies in our visible light correspond to the red end of the rainbow and the high frequencies to the violet end. So, if distance between us and a star is increasing, it demonstrates a shift to the low-frequency end of visible light, "red shift"; if the distance is decreasing, the star demonstrates "blue shift." Our universe is expanding, so almost all stars show "red shift" in their light. Meanwhile, the galaxy Andromeda, adjacent to our home galaxy Milky Way, is approaching us, showing "blue shift." In several billion years these galaxies will collide.

Summary, Terms, Symbols, and Relations

Law of Reflection: the angle of reflection of a wave is equal to the angle of the incidence.

Refraction is bending of the waves due to dependence of the speed of the wave on the position. For sound waves refraction occurs mostly because of the dependence of the speed of sound on the temperature and existence of the wind.

Diffraction is the bending of waves behind the obstacle or behind the opening.

Interference is the addition of the waves. Constructive interference occurs when two waves encounter each other in phase; destructive interference occurs when two waves encounter each other completely out of phase.

Condition of constructive interference:

$$d_1 - d_2 = m\lambda,$$ where m = 0, 1, 2, 3—any integer.

Condition of destructive interference:

$$d_1 - d_2 = m\lambda + \lambda/_2,$$ where m = 0, 1, 2, 3—any integer.

Beats are the phenomenon of the periodic pulsation of the loudness. The frequency of beats is $f_1 - f_2$.

Doppler effect is the apparent shift of a pitch, when distance between source and the listener is changing:

$$\left| f_{received} - f_0 \right| = \frac{f_0}{V} v$$

Questions and Exercises

1. Suppose you are sitting behind a large pillar (or other obstacle) in a concert hall. Will treble and bass notes be blocked in the same way?

2. A window has been opened a distance of 10 cm. What kind of sound from outside will go across the room in a well-defined beam? What kind of sound will be uniformly spread in the room? Roughly what frequency marks the boundary between the two cases?

3. During a windless day the temperature decreases with altitude. How would sound bend in such conditions? What do we call this kind of bending?

4. The wind is blowing from east to west, and the temperature of the air is uniform. How will sound bend as it approaches an observer who is standing to the west of the source?

6. Suppose you listen to sound from two loudspeakers at a distance of 10 m from one and 8 m from another. What are several wavelengths and the corresponding frequencies for which you will experience constructive interference? What are several wavelengths and frequencies for which you will experience destructive interference?

5. Strike a tuning fork against a table and bring it closer to your ear, rotating it slowly. You will hear the oscillations of loudness. How can you explain these oscillations?

7. Two strings produce beats with the frequency of 2 Hz. What would happen to the frequency of beats if you reduce the tension on a string producing lower frequency?

8. If three instruments play together with frequencies 440, 438, and 437 Hz, what beat frequency will result?

9. Two tuning forks produce the frequency of beats 2 Hz. If one tuning fork oscillates at frequency 440 Hz, what is the frequency of the other?

10. Suppose you are standing at the railroad station and the train approaches the station, its whistle blowing. Do you hear the higher or lower pitch in comparison to the situation when the train is at rest? If the train is moving with constant velocity, do you hear the pitch gradually increasing or at a constant, steady pitch?

11. Student A and student B are holding two tuning forks, which demonstrate the frequency of beats 2 Hz. Student A starts to run away and the frequency of beats increases. Which student's tuning fork has a lower frequency?

Chapter 7

ABCs of Music

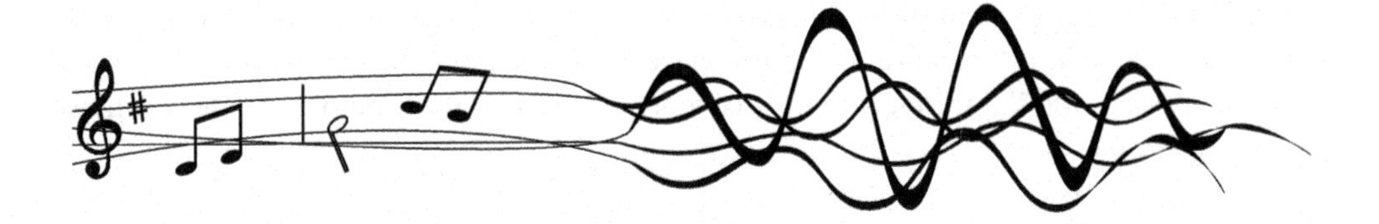

7.1 Some Intro Words

Our goal is to build a bridge between two different areas of human experiences: the fundamental science of waves and oscillations and music. To do this we should be able to speak both languages: of the science and of the music. We already know how to connect our sensation of pitch, loudness, and timbre with frequency, sound intensity level, and the waveform—measurable quantities of Physics.

Music is a very complicated sequence of different sound events: combined sounds connected into patterns, full of pauses, different articulation, and impression. We know very little of how our brains perform a complicated processing of sound signals. We recognize, compare, and even "record" music in our brain, so we can reproduce a melody.

For readers who do not know how to read written music, Appendix B might be helpful.

7.2 Time Element of Music

The second is a very natural basic unit of time measurement for humans. It is commonly explained in connection to the rate of our heartbeat, which is usually slightly less than a second apart. Maybe because of this rate, our other typical reactions and movements also occur in around one second: the typical reaction time of a driver is 1 second, the movements of our arms and legs take usually 1 second, and so on. So, the natural time division 1 second is defined by the construction of our body. If we were small like a fly, our time scale would be significantly shorter.

Note that the limitations in reaction and perception involve the adding of limitations of different organs. Take, for example, the simple reaction of a driver on some sudden obstacle on the road. First, the signal from the driver's eyes should reach the brain, then the brain should process and send the signal to the leg, which then pushes the brake pedal. All these steps require time, resulting in a total reaction time of about 1 second.

The important question is when do we perceive sound events as individual, and when do we start to perceive them like a fast passage with no individual character? The answer is again around one second. If we receive more than 5 events per second, it becomes practically impossible for us to analyze them; if the event (say, a single note) lasts longer than 5 seconds, our brain starts to subdivide it. If the pause in a music piece is less than one fifth of a second, we would not even notice it, and if it is longer than 5 seconds, we consider it an interruption between two separate music pieces.

The fast passages, played on piano with rate of up to 10–15 notes per second, are not considered by our brain as 10 or 15 independent musical events. Our brain considers the whole sequence of notes as one signal. The progression of major events in a musical piece, based on significantly different chords, which is called **harmonic tempo** or harmonic rhythm, is again around 1 second. For example, in Prelude no. 1 in C major, J. S. Bach changes chords after each stream of 16 notes. Another example, "The Minute Waltz" of F. Chopin, also shows that our brain is not able to process information too fast. This music piece, by a legend, was created to be played in one minute. But it contains 420 notes. It is physically possible to play it in such tempo, but the music loses any sense, becoming a total mess. As a result, it takes usually 2 minutes to play this piece in a reasonable articulation.

After this preliminary discussion we should finally give a definition of **tempo**. In musical terminology **tempo** (Italian for *time*) is the speed of a given music piece. The tempo of a piece will typically be written at the start of a piece of music, and in modern Western music it is usually indicated in **beats per minute (BPM)**. This means that a particular note value (for example, a quarter note) is specified as the beat, and the marking indicates that a certain number of these beats must be played per minute. The greater the tempo, the larger the number of beats that must be played in a minute.

BPMs became commonly used in disco by DJs. In modern music beats measured are often not the quarter notes, but drum beats, whichever is fastest. This allows us to reach higher BPM values by increasing the number of drum beats, without increasing the tempo of the actual notes. Typical BPM values for modern music are 120–125 for House music, 160–180 for Jungle music, and Hardcore techno exceeds 180 BPM.

The use of extreme BPMs was very common in the fast bebop jazz of the 1940s and 1950s. It was achieved with very fast drum patterns at the same underlying tempo. A jazz standard "Cherokee" was often performed with a tempo exceeding 368 BPMs.

Some musical pieces do not have a mathematical time indication. In classical music it is customary to describe the tempo of a piece by one or more words. Most of these words are in Italian because many of the most important composers of the seventeenth century were Italian, and this period was when tempo indications were first used extensively. The code words for tempo are shown in Table 7.1. Note that each word allows a lot of freedom for a performer; it is not a strict order, but rather a recommendation.

Table. 7.1. Correspondence between Italian titles for tempo and metronome settings.

Tempo Marking	Translation	Metronome Setting
Largo	Broad	40-70
Larghetto		70-100
Adagio	At ease	100-128
Andante	Going	128-156
Allegro	Gay	156-184
Presto	Fast	184-208

In the second half of the nineteenth century, J. N. Mälzel invented the metronome—a device that produces regular, metrical clicks, settable in BPMs. Several composers started to use metronomes immediately, writing metronomic indications for their music pieces. Some of these marks are not very useful; for instance, Schumann often gave marks for such a high tempo that it was impossible to perform. Now metronomes are highly precise electronic devices, widely used in music sequencers—devices or software for mixing and processing digital audio. A good performer usually does not mechanically follow the metronome tempo, changing the rate slowly, using the tempo as another tool to increase the impression.

Most of us can readily identify time patterns of strong and weak beats in music. This is the **rhythm**. Here we use the word "beat" to mean a regularly spaced series of accented notes. These patterns repeat throughout the composition.

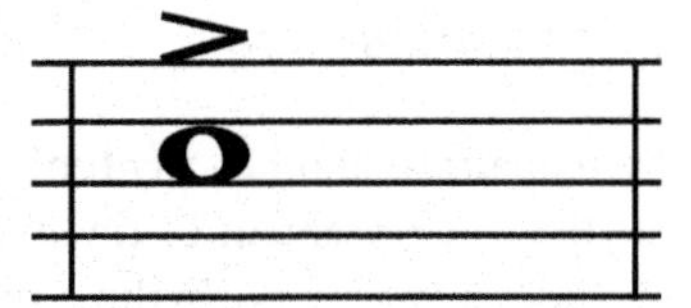

Figure. 7.1. Example of the accent mark often used in written music.

Music from certain cultures (Africa, China, Japan) uses very complex rhythmic patterns, often with several rhythms intertwined and played by different instruments. These cultures use a lot of percussion instruments, so the rhythm becomes the main feature of the music. The combination of sustained harmonies of Western music and complex rhythms from other cultures is a pretty challenging problem both for performers and for audiences; it contains too much musical information to be easily processed by the brain of an average listener.

By contrast, European cultures use a lot of strings and wind instruments; the primary feature is the developing of complex harmonies, so the rhythmic patterns in much of European classical music are pretty basic and clear-cut. The easiest rhythm is $\frac{2}{4}$, which shows a pattern SwSwS …, where S stands for a strong beat and w for a weak beat. Because this rhythm has one strong, one weak beat per measure, and we have two legs, the music with this rhythm has obvious marching character. The $\frac{4}{4}$ rhythm, such as SwwwSwww … also could be used as a march because the human brain subdivides it into group 2+2, and the second weak beat could serve as additional strong one. The rhythm of $\frac{3}{4}$ SwwSww … is widely used for dancing and corresponds to a classical waltz. The odd number of beats per measure requires special patterns of dance, with two big steps and a third, smaller one, which returns the dancer to the initial position.

Basic time units are arranged in groups that usually correspond to vertical bar lines in the written score (called measures). Strong beats can be indicated by accent marks (Fig. 7.1). The way these units are arranged is known as **meter**. The term meter was inherited initially from poetry, where it describes the number of lines in a verse, the number of syllables in each line, and their arrangement. Meter is shown at the beginning of each music piece and (if the meter changes) in the middle, with signs such as $\frac{2}{4}$ (meaning two quarter-note beats per measure) or $\frac{3}{4}$ (three quarter-note beats per measure). The bar lines make it easier for performers to keep track of their place in the music. But this advantage exists only for pieces with fixed meter. A lot of non-European music (for example, original pieces of Honkyoku for the Japanese bamboo flute) is written with irregular meter and does not use bar lines.

Rhythms can involve the division and subdivision of time units into two or three parts, and so are written in what we call it duple or triple meters. Rhythms that divide each of two main beats into three subbeats call for a compound meter such as $\frac{6}{8}$. The $\frac{3}{4}$ meter has also six eighth-notes per measure, but they are arranged into three groups of two.

The rhythms consisting of the rarer patterns $\frac{5}{4}$ and even $\frac{7}{4}$ are perceived as alternating subgroups 2+3 and 3+4; the human brain breaks groups of five or more things into smaller units. Although these rhythms are not so common, there are famous pieces that use so exotic a pattern. For $\frac{5}{4}$ rhythm, the most popular became the compositions of Dave Brubeck ("Take Five") and Andrew Lloyd Webber ("Everything's Alright" from *Jesus Christ Superstar*). The $\frac{7}{4}$ rhythm is used in the composition "Money" by Pink Floyd, "Unsquare Dance" by Dave Brubeck, and "The Arrest" in *Jesus Christ Superstar.*

Syncopation, another rhythmic device, refers to a disruption or break in the regular flow of rhythm—a placement of rhythmic stresses or accents where they wouldn't normally occur. It includes a variety of rhythms that are in some way unexpected in that they deviate from the strict succession of regularly spaced strong and weak but also powerful beats in a meter. These include a stress on a normally unstressed beat or a rest where one would normally be stressed.

Syncopation has been an important element of musical composition since at least the Middle Ages. Bach, Handel, and Haydn used syncopated rhythms as an inherent part of their compositions.

Syncopation is used in many musical styles, and it is fundamental in African-derived styles such as jazz, ragtime, blues, rap, and reggae. All dance music uses a lot of syncopation. Hungarian Csárdás song-dances are always syncopated. The Scotch snaps of Scotland also feature syncopation.

Several simple types of syncopation can be considered:

- missed-beat syncopation (Fig. 7.2a), which just involves the addition of a rest. Silence in this situation substitutes for the expected note and can even fall on the strong beat. It creates additional "tension" in musical pieces.
- even-note syncopation. In meters with even numbers of beats ($\frac{2}{4}$, $\frac{4}{4}$, etc.), the stress normally falls on the odd-numbered beats. If the even-numbered beats are stressed instead, the rhythm is syncopated.

- off-beat syncopation, when stress is shifted by less than a whole beat. See Fig. 7.2b where the accent is shifted back by an eighth note; meanwhile, the expected beat should be like it appears in Fig. 7.2c.

Another interesting tool to add flavor is the **hemiola**. This word in Greek means "one and a half," or the ratio 3:2. In a music piece that normally follows the SwwSww pattern, it appears to occasionally insert the SwSw element, while the tempo of subbeats is kept unchanged. The hemiola is commonly used in classical music, especially in Baroque and Flamenco dance forms, as well as in nineteenth century scherzos. Composers who have used the hemiola particularly extensively include G. F. Handel, C. M. von Weber, J. Brahms, and P. I. Tchaikovsky. A regular alternation 2x3 + 3x2 continuing measure after measure all throughout the music piece is used in the modern composition of Matt Savage Trio "Infected by Hemiola." The last piece can be used to study this interesting device for changing the rhythm.

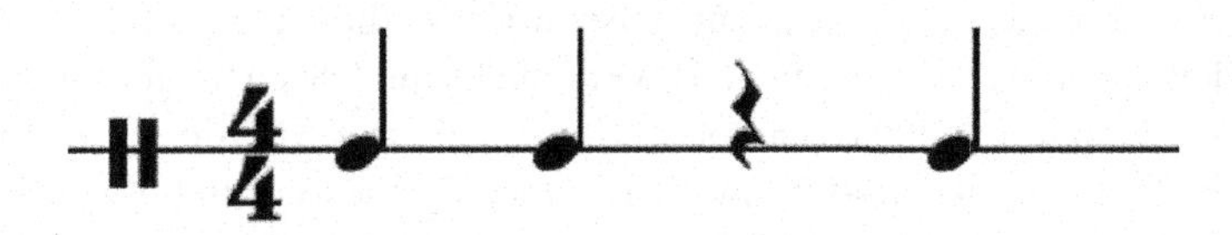

Fig. 7.2a. Missed-beat syncopation: addition of rest.

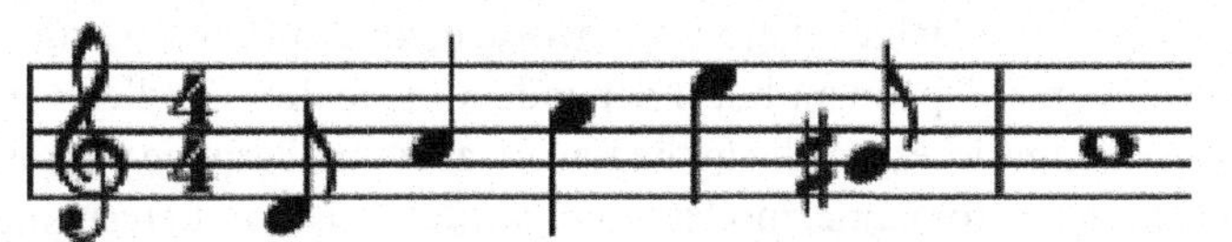

Fig. 7.2b. Off-beat syncopation: the accent in this figure is shifted to the eighth note instead of as it is shown in Fig. 7.2c below.

Fig. 7.2c

7.3 Horizontal and Vertical Structure of Music

A **melody**, also **tune**, **voice**, or **line**, is a succession of musical pitches that is perceived as a single entity. In its most literal sense, a melody is a combination of pitch and rhythm. Melody gives a horizontal structure of music, progressing from one note to another. It can be played on any instrument that has definite pitch. Melodies often consist of one or more musical phrases or motifs, and are usually repeated throughout a piece in various forms. Simple vocal melodies seldom cover a range of more than one octave.

Melody is used in different musical styles in various ways. Jazz musicians use the word **lead** to refer to the main melody, which is used as a starting point for improvisation. Rock music and folk music often pick one or two melodies and stick with them throughout the piece. In Western classical music, composers often introduce an initial melody, and then create variations. Often, melodies are constructed from motifs or short melodic fragments, such as the famous opening of Beethoven's Fifth Symphony. Richard Wagner popularized the concept of a motif or melody associated with a certain idea, person, or place, which he called leitmotif.

A harmony is the use of simultaneous pitches. The study of harmony involves chords and their construction and the principles of connection that govern them. Harmony is often referred to as the "vertical" aspect of music. Harmony can exist purely for the support of the melody, working like a decoration, "dressing" the pure melody with fuller sound. Such usage of harmony is pretty common in popular music. A good example of this is Barbershop music, where four voices sing a capella. Each of the four parts has its own role: generally, the lead sings the melody, the tenor harmonizes above the melody, the bass sings the lowest harmonizing notes, and the baritone completes the chord, usually below the lead.

Much more difficult and yet interesting is to create music for several performers (voices), which is different and highly structured both vertically and horizontally, and where each performer has his or her own melody. Such a relationship between voices of several different instruments that are harmonically interdependent but independent in rhythm is called **counterpoint**. It is widely used in classical music, especially in the Renaissance and Baroque periods. In many types of music, notably baroque, romantic, modern, and jazz, chords are often augmented with "tensions." A tension is an additional chord member that creates a relatively dissonant (unpleasant, not rough) sound, together with the bass note. Any harmony to sound pleasant and interesting for our ears should contain balanced amounts of "tense" and "relaxed" chords.

7.4 Articulation

Articulation, the word that we know has different meanings, refers in music to the manner in which successive notes are joined to one another by a performer.

There are many types of articulation, with each having a different effect on how the note is played. Each articulation is represented by a different symbol placed above or below the note. The typical types of articulation are shown in Fig. 7.3:

Fig. 7.3. Examples of articulation: legato, staccato, staccatissimo.

- **staccato**, signifies a note of shortened duration, separated from the note that may follow by silence. Each note is released quickly with no sound during the time that would ordinarily be occupied by that note;
- **staccatissimo**, indicates that the notes are to be played extremely separated and distinct, a superlative staccato;
- **legato**, indicates that musical notes are played or sung smoothly and connected. That is, in transitioning from note to note, there should be no intervening silence.

Professional musicians use many gradations in between staccato and legato. The importance of particular gradations depends on the instrument. For example, on a piano it is possible to stress strong beats by playing them louder. But on the harpsichord, articulation is the only form of stress available, so a performer should understand clearly the interchange of sounds and silences in a piece.

Stress in articulation influences not only the silent gaps between notes but also the apparent loudness of the notes. Fig. 7.4 shows that a note should be sustained for several tenths of a second until the ear is able to judge its real loudness. Notes that last less than a few tenths of second seem less loud than does continuous sound of the same intensity.

The gaps between notes are also not exactly silent because of room reverberation, as we will discuss in Chapter 15. If reverberation time is longer than a silence between staccato notes, the overlapping of sounds can create the effect of "muddiness."

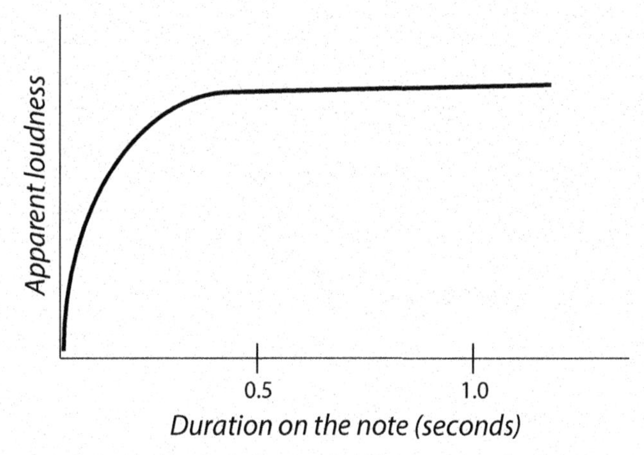

Fig. 7.4. Dependence of the perceived loudness on the duration of the note.

7.5 Scale in Equal Temperament

Any octave on a piano contains a lot of recognizable pitches. So, the question is how we should divide this octave into steps (notes). Many instruments (such as trombone with a gradually changing resonator, or any fretless string instrument) allow us to play all the possible pitches. But pianos and guitars are committed to using only 12 pitches per octave. Note that 12 steps per octave is a choice for European music only. For example, Indian classical music uses a seven-note system. Microtonal music, with roots in Western music, also uses widely different systems of temperament. The composer Harry Partch made custom musical instruments to play compositions that employed a 43-note system, and the American jazz vibraphonist Emil Richards experimented with such scales in his "Microtonal Blues Band" in the 1970s.

Scale is a sequence of allowed musical pitches in ascending and descending order. Most commonly, the notes of a scale belong to a single key (for example, C major), thus providing material for or being used to represent a melody and/or harmony.

Chromatic scale (Fig. 7.5) consists of all 12 allowed pitches per octave, or 12 **semitones**. So, two adjoining white keys on a piano are tuned to create a semitone. Two adjoining white keys that have a black key between are a **tone**.

An **equal temperament** is a musical temperament, or a system of tuning, in which every pair

Fig. 7.5. Chromatic scale in equal temperament.

of adjacent notes has an identical frequency ratio. This means that any two adjacent semitones, no matter in what range of the piano, must create the same ratio: such as A and $A^{\#}$ or C and $C^{\#}$.

Let us consider how to create an equal temperament in an example:

We decided to tune a piano in such a way that the frequencies of adjacent notes create the same ratio, say x. We start our tuning from A_4, which has frequency 440 Hz. Then the frequency of $A_4^{\#}$ will be $440 Hz \cdot x$. But the frequency of the next note, B_4, should be bigger than the frequency of $A_4^{\#}$ also x times: $440\ Hz \cdot x \cdot x = 440\ Hz \cdot x^2$. The frequency of next note, C_4, is another x times bigger than the frequency of B_4, and so on. For the frequency of each of the following notes, we should multiply the frequency of previous one by x. After the 12 steps corresponding to an octave, we come to A_5, whose frequency is equal, following our recurring rule, to 440 $Hz*x^{12}$. But we also know that A_4 and A_5, being one octave apart, have a ratio of frequency 2. As a result:

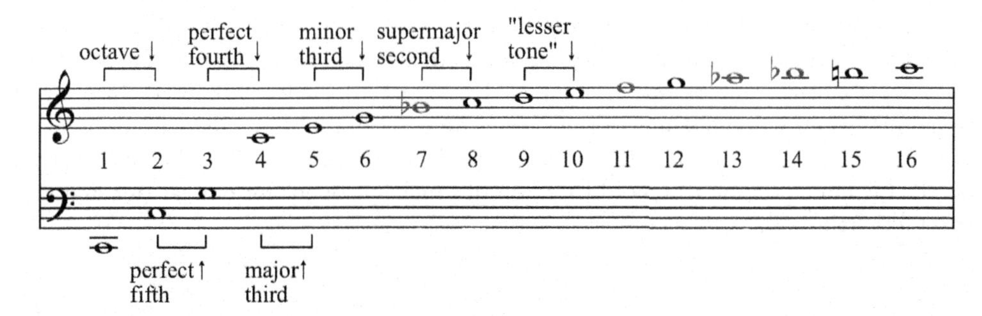

Fig. 7.6. Harmonic series based on fundamental C_2.

$$440\,Hz \cdot x^{12} = 440\ Hz \times 2;$$

$$x^{12} = 2$$

$$x = \sqrt[12]{2}$$

From the last equation, x = 1.05945. This is the ratio of two notes with a difference in pitch of one semitone.

The very important advantage of such a temperament is that the value of x does not depend on the frequency of the note from which we start our tuning: the frequency of A_4 disappeared from our calculation.

Example 7.1. What is the frequency of the note two semitones higher than 440 Hz?

Answer:

The frequency of the note two semitones higher than 440 Hz can be written as:

$$440\ Hz \times 1.059 \times 1.059 = 493.9\ Hz.$$

7.6 Musical Intervals

The perceived spacing between two pitches is called an **interval**. Intervals can be vertical or **harmonic** if the two notes sound simultaneously, or horizontal, linear, or **melodic** if they sound successively. Three or more notes sounding together form a **chord** containing the combination of several intervals.

An octave, for example, is the interval of 12 semitones. Well-trained musicians are able to recognize each of several other intervals for they have distinctive aural qualities. The intervals C_4–G_4 and A_4–E_5, for instance, involving different pitches have both the same name (perfect fifth), because they sound alike in the same sense as two octaves sound alike. For purposes of musical harmony, the interval is more important than the actual value of individual pitches.

The common names for the intervals (third, fifth, octave) are just a way of telling how many white keys they span on the piano. Because a keyboard also has black keys, there are adjectives added to these names (minor, major, augmented, diminished). Table 7.2 shows the naming of common intervals with the number of semitones for each. It also shows the ratio of frequencies of higher and lower notes in the interval. Note the closer this ratio is to the ratio of small integers, the more pleasant the chord based on this interval sounds for our ear (for example, perfect fifth, ratio 3:2).

Table. 7.2

Interval Name	Number of Semitones	Ratio of Frequencies
Unison	0	1
Minor Second	1	1.059
Major Second	2	1.12
Minor Third	3	1.19
Major Third	4	1.26
Perfect Fourth	5	1.33
Tritone or Augmented Fourth or Diminished Fifth	6	1.41
Perfect Fifth	7	1.49
Octave	12	2

7.7 The Harmonic Series

Let us perform an easy experiment with any keyboard instrument, such as a piano. Pick a note, say A_2, which has frequency 110 Hz and play a sequence of pitches 220Hz (A_3), 330Hz (E_4), 440Hz (A_4), 550 Hz ($C_5^\#$), 660 Hz (E_5). … We play a group of notes whose frequencies are simple multipliers of the initial note, A_2. The arpeggio based on this sequence sounds very pleasant and, in some way, predictable to our ear. It seems to us that we can actually predict each next note.

The group of frequencies that builds as simple multiplication of basic frequency is called a **harmonic series.** The members of such a group are called **harmonics** and the lowest harmonic (in considered example A_2) is called the **fundamental.** Other members are called second harmonic, third harmonic, and so on. Fig. 7.6 shows the harmonic series built on fundamental C_2.

Let us analyze this sequence in more detail (see, for example, Fig. 7.6). The ratio of frequencies of the first two notes creates ratio 2:1, and it's an octave; $f_3/f_2 = 3:2$, perfect fifth; $f_4/f_3 = 4:3$, perfect fourth; $f_5/f_4 = 5:4$ close to major third; $f_6/f_5 = 6/5$ close to minor third. The next expected member of this sequence, 770 Hz, falls between notes $F_5^\#$ and G_6. If either of them is played, it sounds strange and off the harmonic sequence. So, the six first notes in this example of harmonic series sound **very close to intervals of equal temperament.** The seventh harmonic in this example falls in between the pitches of the piano keyboard. Yet this harmonic is a legitimate member of harmonic series and is musically important.

Example 7.2. The fundamental of some harmonic series is 200 Hz. What is the fifth harmonic?

Answer:

The fifth harmonic of this series is 200 Hz x 5 = 1000 Hz.

Example 7.3. The third harmonic of some series is 450 Hz. What is the seventh harmonic?

Answer:

First we should find the fundamental of the given series. The fundamental is

$\frac{450Hz}{3} = 150Hz$. The seventh harmonic is $150Hz \cdot 7 = 1050Hz$.

Example 7.4. Pitches 450 Hz and 600 Hz belong to the same harmonic series. What is the fundamental of this harmonic series?

Answer:

The fundamental of the harmonic series is the biggest common divisor of two frequencies, 450 Hz and 600 Hz: 150 Hz.

Summary, Terms, and Relations

Tempo is the speed of a given music piece.

Meter shows the way the basic time units are arranged in groups, usually corresponding to vertical bar lines in the written score.

Rhythm of music is a recurring time pattern of strong and weak beats.

An **equal temperament** is a musical temperament, or a system of tuning, in which every pair of adjacent notes has an identical frequency ratio.

The ratio of frequencies of two notes with one semitone difference in pitch in European equal temperament is equal to 1.059.

Questions and Exercises

1. Using Figure 5 from Appendix B, find the names of notes that are nearest to the frequencies (a) 270 Hz, (b) 1000 Hz, and (c) 560 Hz.

2. What intervals are formed by the following pairs of notes: (a) E_4/G_4; (b) C_4/D_4; and (c) C_4/E_4? For reference, use Fig. 5 of Appendix B.

3. What is the ratio of frequencies for the intervals (a) major second, (b) major third, (c) tritone, and (d) perfect fifth?

4. If A_5 is the fifth harmonic in a series, what is the fundamental? For reference, use Fig. 5 of Appendix B.

6. What is the closest musical interval to the one created by the third and fourth harmonics of the harmonic series?

5. A harmonic series starts from A_2 (110 Hz). What are the names of the first five notes of this harmonic series? What is the number of the first harmonic that falls in between the keys of the piano keyboard? For reference, use Fig. 5 of Appendix B.

7. What note is an octave above A_3? One perfect fifth below E_5? A major third above B_3? For reference, use Fig. 5 of Appendix B.

8. What frequency corresponds to each of the following notes in the equal-tempered chromatic scale: (a) A_1; (b) B_2; (c) C_3; (d) D_4; and (e) E_5? For reference, use Fig. 5 of Appendix B.

9. If C_4 is the fourth member of the harmonic series, what is the name of the fifth harmonic?

10. What are the three intervals in the chord $C_4/E_4/G_4$ (major triad)? What are the three intervals in $D_4/F_4/A_4$ (minor triad)?

Chapter 8
Perception of Sound

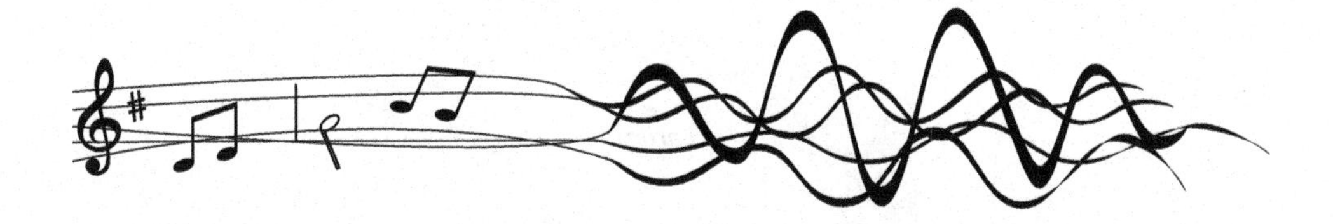

8.1 The Human Ear

In a previous chapter we discussed the audible frequencies and bearable amplitudes of human hearing, as well as the measurable unit, which is the closest to human perception, the decibel.

From everyday experience we know that different frequencies can be felt differently for each person; human perception is subjective. Sounds that are low in amplitude can be perceived as loud, depending on frequency.

In this chapter we will discuss the properties of human hearing because how we perceive the sound depends not only on a source but also on a receiver, the human ear.

Our ear is a sophisticated device, which not only perceives a wide range of frequencies and amplitudes but also transforms the elastic sound wave into series of electromagnetic pulses, "readable" for our brain. Our brain does not understand elastic waves, it processes electromagnetic pulses. That's right: our ear is a **converter**.

All parts of our ear are pretty small and hidden under the bones of our temple to protect them from external damage (Fig. 8.1).

The **outer ear** consists of the **pinna** and **auditory canal**. The pinna is the visible, outer part of ear, different in each individual. The inner part of pinna, approximately 3 cm across, works like an antenna, helping to guide short-frequency sound toward the eardrum. The ear canal itself is about 1 cm across and 3 cm long, acting like a resonator, boosting hearing sensitivity in the frequency range 2000–5000 Hz. The outer ear ends at the very superficial layer of the tympanic membrane, commonly called the eardrum. The eardrum consists of fine and delicate fibrous

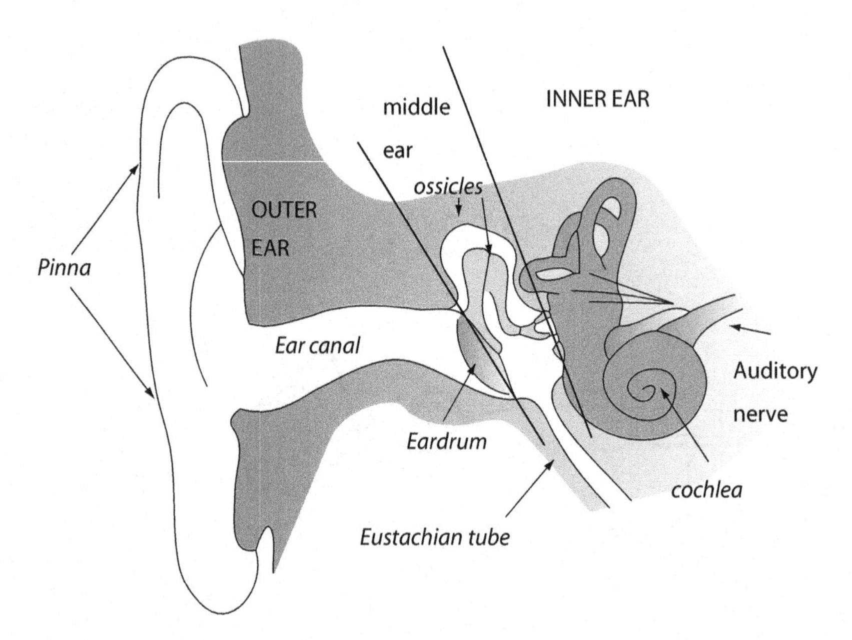

Fig. 8.1. Internal structure of human ear.

tissue. Although the pinna and outer canal help to direct sound toward the eardrum, they also protect the delicate eardrum from outer damage. The eardrum behaves like a drum membrane under the influence of changing pressure, bulging in and out.

The middle ear begins with the above-mentioned eardrum, which is kept taut by a tensor tympani muscle. After the eardrum, there is a chamber that is connected to the throat and, as a result, to the outer world by the Eustachian tube.

The Eustachian tube equalizes pressure on both sides of the eardrum, which is important when the outer pressure is changing due to weather or when climbing a mountain. Unequal pressures distend the eardrum, so the hearing ability decreases, creating of sensation of "blocked" ears, especially for the low-frequency range. The Eustachian tube is approximately 35 mm long and is normally closed, opening only to let a small amount of air in to equalize pressure.

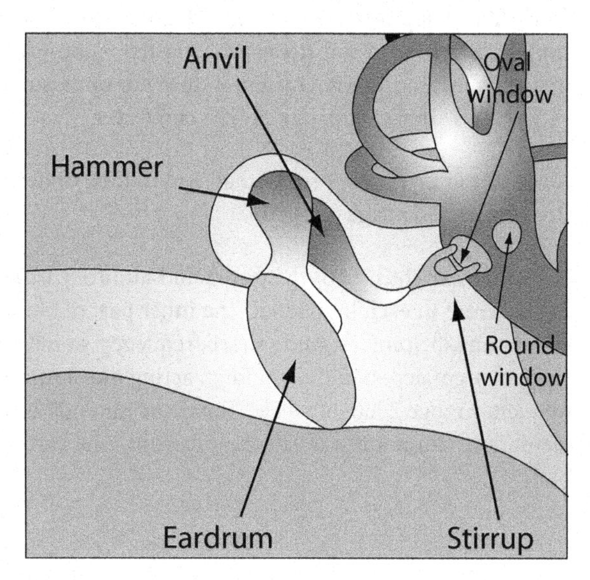

Fig. 8.2. Ossicles: the main part of the middle ear.

Other important parts of the middle ear are the **ossicles**—three little bones, shaped like a hammer, an anvil, and a stirrup (see Fig. 8.2). The handle of the hammer is secured on the eardrum, and the stirrup "knocks" on the oval window that separates the middle and inner ear.

The ossicles play a very significant role in the hearing process. Together they act as the lever, which changes with very small pressure, exerted by a sound wave on the eardrum, into pressure

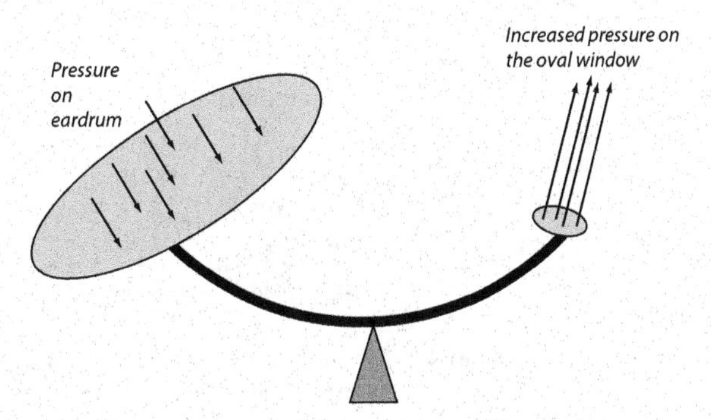

Fig. 8.3. Because of the action of the ossicles, the small pressure on the eardrum is enhanced on the oval window.

up to 30 times higher on the oval window of the inner e,ar (Fig. 8.3). The lever action of ossicles provides the multiplication of force of about 1.5 times, and another 20 times of difference in pressure comes from the difference in the areas of the eardrum and oval window.

Ossicles are not rigidly connected to each other because their other important role is to protect the area of the inner ear from very loud sounds and sudden pressure changes. Loud noise triggers two sets of muscles, one tightens the eardrum, and the other pulls the stirrup away from the oval window. This response to loud sounds is called the **acoustic reflex**. This phenomenon occurs when the ear is exposed to sound of 90 dB and higher. It is worth noting that this reflex takes place only after 30 or 40 ms from the time the sound overload happens, and full protection requires another 100 ms or so. This is the reason why it can be too late to protect the ear in the case of a loud, abrupt sound, such as a gunshot or explosion.

The **inner ear** starts right after the oval window and contains **semicircular canals** and the **cochlea**. The semicircular canals are three semicircular interconnected tubes perpendicular to each other and do not contribute to hearing ability. They are filled with the fluid endolymph and work as sensors of balance.

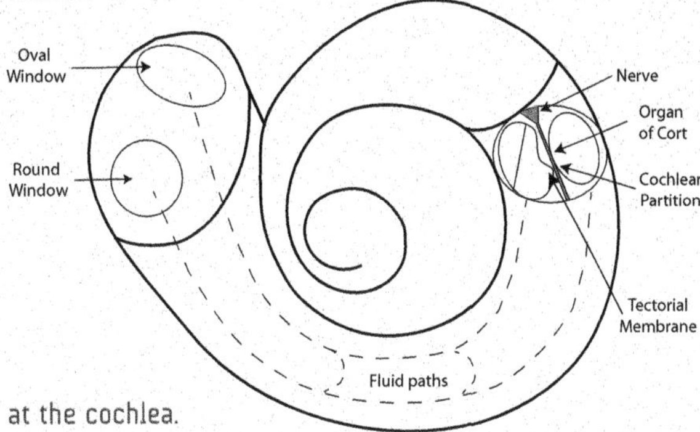

Fig. 8.4. A closer look at the cochlea.

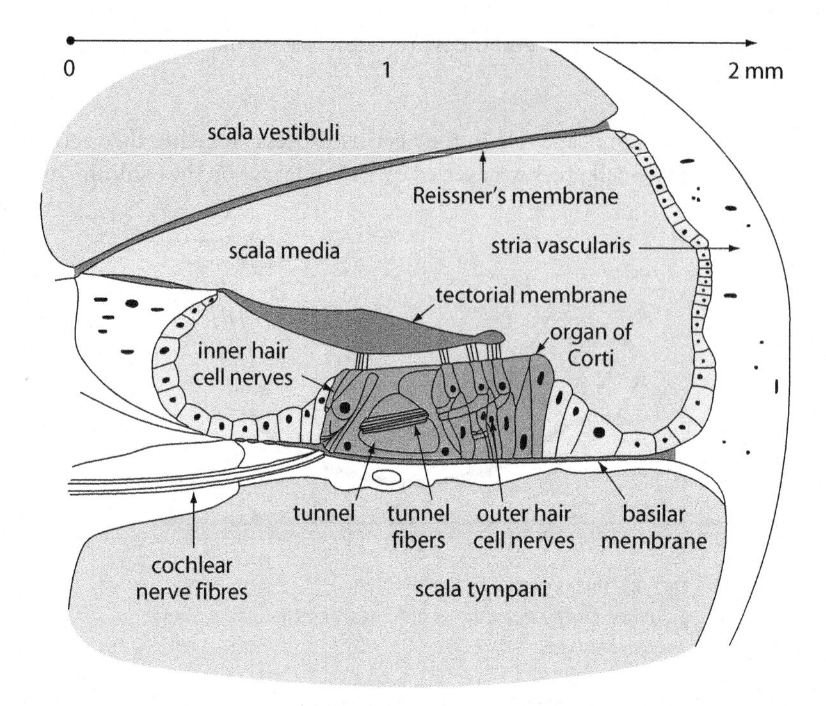

Fig. 8.5. The cross-section of cochlea.

The cochlea is a hollow chamber of a bone that looks like a little snail (Fig. 8.4) and contains all the mechanisms for the transformation of pressure waves into electromagnetic pulses, proper for our brain. Uncoiled, the cochlea is only 3–3.5 cm long.

A cross-section of a cochlea is shown in Fig. 8.5. There are three distinct chambers that run the entire length: the scala vestibuli, the scala tympani, and the cochlear duct. Two different fluids fill the chambers of cochlea: perilymph in scala vestibuli and endolymph in cochlear ducts. These two fluids have different densities and different viscosities. The two liquids are separated

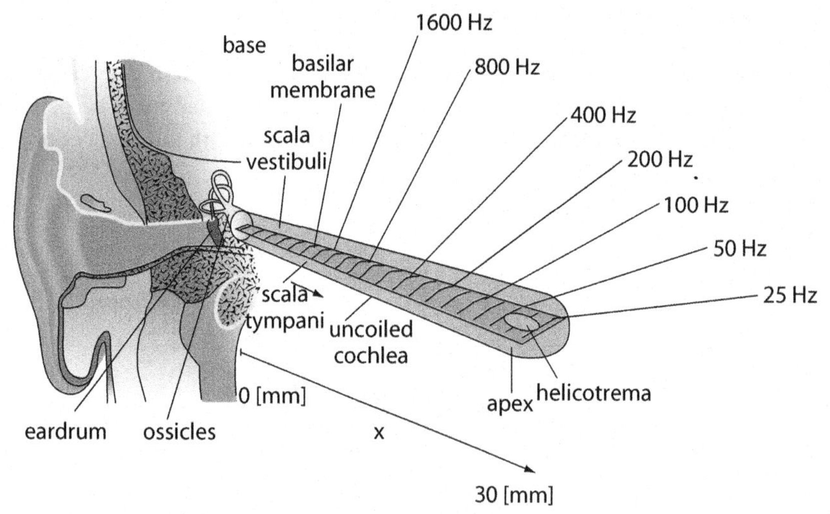

Fig. 8.6. "Uncoiled" cochlea. The typical frequencies excite at different distances from the oval window are shown.

by two membranes: **Reissner** membrane, which is very thin and separates the scala vestibuli from the cochlear duct; and **basilar** membrane, a main structural element that separates the scala media from the scala tympani and determines the mechanical wave propagation properties of the cochlea.

Riding on the basilar membrane is the delicate organ of Corti, a gelatinous mass of about 3 cm in length, in which sensory hair cells are powered by the potential difference between the perilymph and the endolymph. It contains several rows of tiny hairy cells, which have 15,000–20,000 auditory nerve receptors attached to them. Each receptor has its own hair cell, and each hair cell has many hairs that bend when the basilar membrane responds to a sound. The bending of hairs activates the cells, which send the excitation into neurons of the auditory nerve.

Let us consider cochlea uncoiled and simplified, which will help us to understand the vibrations of the basilar membrane (Fig. 8.6). The cochlea looks like a cylinder divided into two parts by the basilar membrane. At the thicker end of the cylinder, there are oval and round windows, each closed by a delicate membrane, and near the far end of the basilar membrane is a small hole, the helicotrema, connecting the two parts. The basilar membrane ends short of the narrow end of a cylinder, so fluid can transmit waves around the end of the membrane.

The stirrup vibrates, knocking into oval window; the waves in fluid are transmitted down the scala vestibuli, creating ripples in the basilar membrane. The basilar membrane is not uniform: it is stiff and thin at the oval window and "floppier" at the far end. As a result, high frequencies create the biggest amplitude close to the oval window; meanwhile, low tones set into vibration the longer column of fluid and floppier end. So in the cochlea, our ear already starts to analyze different pitches.

To conclude: sound waves travel through the ear canal and set the eardrum into vibration, causing the oscillations in the middle ear. The stirrup, vibrating against the oval window, causes vibrations in the cochlea and basilar membrane. As a result, bent hairs excite the hair cells, which transmit electrical pulses into our brain.

It is worth noting that some sounds can be heard through the vibration of the skull's bones, which also reach the inner ear. Such sounds as the clicking of teeth are transmitted almost entirely through the bones. During speaking we hear both sounds: through the bones and through the ear. This explains why hearing our own recorded voice seems unfamiliar and unnatural. The recorded sound is a result of the airborne vibration detected by a microphone; meanwhile, we always hear both components.

8.2 Limits of Perception

The human ear can perceive sound waves between 17 Hz and 17 kHz, but the sensitivity of the ear is not uniform. We are the most sensitive to sounds from 2 to 5 kHz. A sound of 0 dB is used as the softest possible sound. But it requires really ideal and special conditions to hear 0 dB. Most people require levels of 15–20 dB for high frequencies and even higher levels (above 30 dB) for low frequencies. The record for the lowest pitch heard is 12 Hz under special conditions; sounds below 10 Hz can be felt by bone vibrations.

The upper limit of audible frequencies depends on age. For a young person the range is 17–18 kHz, then it gradually decreases once into adulthood, and by retirement age, it is around 12 kHz for women and 5 kHz for men. This still leaves us a possibility to enjoy music because the uppermost keys of a piano are 4.5 kHz.

If you decide to perform an experiment with infrasound—frequencies below the audible range, which can be felt only by body vibrations—you better avoid the range 7–10 Hz, which may cause nausea.

8.3 Just Noticeable Differences (JND)

The human ear has its limits of resolution, not only for loudness and frequency range. Two very close frequencies will sound indistinguishable for us as well as the sound of two sources not so different in loudness.

Perception of the human ear involves not only receiving information but also transmitting it inside the ear and nerves and processing it. As a result, the information that we finally get is filtered by the properties of our "receiver." The science that establishes and studies the connections between physical characteristics of sound and actual perception of humans is a part of psychophysics.

One of the questions being considered by psychophysics is: Among two audible sounds, how much must one differ from another so we can detect the difference? This question is not only a question of acoustics but of any science that deals with human perception: optics, physical therapy, culinary.

Psychophysics answers this question with a standard procedure of the determination of **just noticeable differences** (**JNDs**, sometimes called **difference limens**).

We will use two sources of sound sine waves to determine the minimum detectable changes. Let us start with the sound level. We compare two sources, A and B, both sine waves with only slightly different sound intensity levels. We set one source (A) at fixed level (say, 50 dB) but change the loudness of B with time. The subject of our experiment hears A and B alternatively, and is required to tell after each set which one sounds louder. It is obvious that if these two sources differ by, say, 20 dB, everybody will make a correct judgment. But if both levels are the same, and listeners are not allowed to pick "the same" as an option, 50% of listeners will pick source A and 50%, source B.

As the level of a source B gradually increases, more and more listeners pick it as the louder one (Fig. 8.7). A typical result for a sine wave of frequency 1000 Hz is $SIL_B = 50.9$ dB for 75% correct choices, indicating that half of choices are random, but another half is based on a difference actually perceived. We say that the JND in sound level is 0.9 dB for 1000 Hz sine wave at 50 dB.

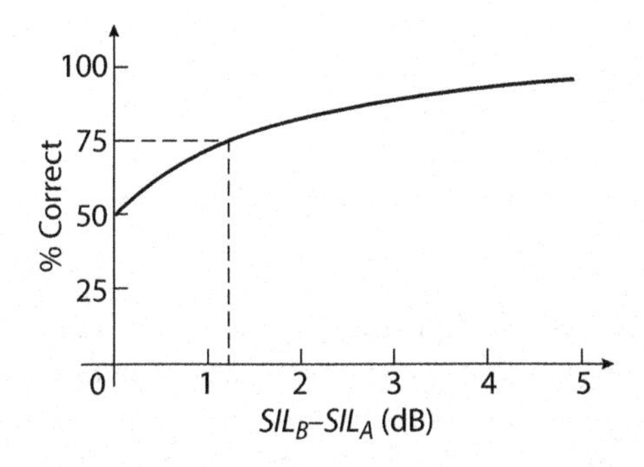

Fig. 8.7. How we define the just noticeable differences (JND).
Source: Musical Acoustics by Donald Hall, 2002 Cengage Learning, Inc.

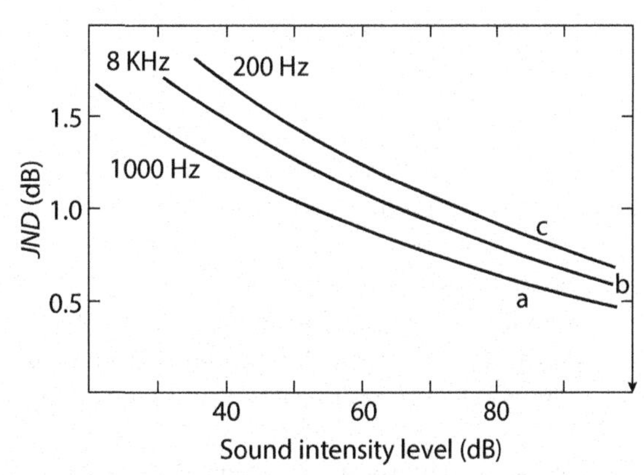

Figs. 8.8a, b, and c. Just noticeable differences (JND) in loudness for (a) 1000 Hz,
Adapted from Musical Acoustics by Donald Hall, 2002 Cengage Learning, Inc.

This procedure can be repeated with the fixed source A reset for other sound levels to establish corresponding JND for each level. When we collect all these data, the dependence of JNDs for 1000 Hz sine wave is obtained (Fig. 8.8a). After this we repeat the entire experiment for other frequencies (Figs. 8.8b, c).

The JNDs for all frequencies important for musical interest (from 500 Hz to 4000 Hz) are approximately the same as for 1000 Hz: ½ to 1 dB. Note that this change in sound level requires change in intensity by 15–30%. For soft sound, around 30 dB, it requires an even bigger change in intensity, 60%, before you notice the difference.

The determination of JNDs in frequency requires the analogous procedure. The listeners are exposed to alternating tones of equal loudness with one frequency kept constant. The frequency of the second source is varied to find what minimum frequency difference is needed to define reliably which of these two tones has a higher pitch. This is repeated for other values of fixed frequency to get the curve in Fig. 8.9. For all frequencies below 1000 Hz, it is safe to say that JNDs are 1 Hz and increase after this. At 3 kHz, the JND rises rapidly, and our pitch judgment becomes really poor. Above 10 kHz, the ability to hear differences in pitch vanishes. For low frequencies (around 50 Hz) the JNDs may be smaller than ½ Hz.

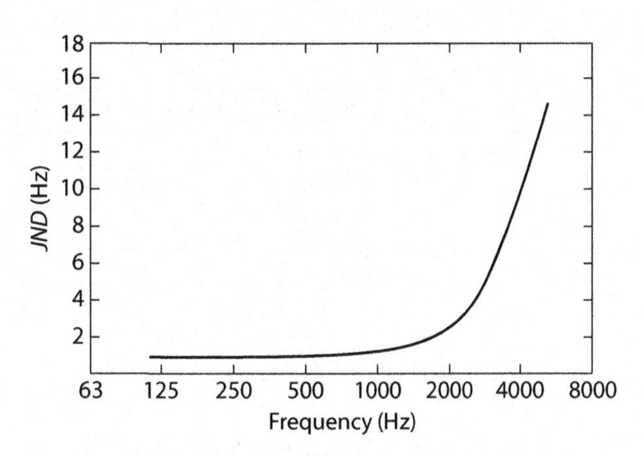

Fig. 8.9. JNDs in frequency (pitch). Note the increasing of JNDs for high pitch above 2000 Hz.
Adapted from Musical Acoustics by Donald Hall, 2002 Cengage Learning, Inc.

8.4 Variables of Physics vs Variables of Psychophysics

We already know the characteristics of a steady tone: amplitude (intensity and SIL are all connected), frequency of vibration of a source, and waveform. All these variables can be measured with corresponding devices.

Psychophysics has many more characteristics of sound based on human perception. It includes not only loudness and pitch but also fullness, smoothness, the character of an instrument, brilliance, and maybe much more. So, the character of an instrument (good/bad violin or good/bad trumpet) we can try to pack into a category called timbre. It describes tone quality or tone color.

Unfortunately we cannot claim that pitch is defined by frequency, loudness by amplitude, and timbre by a waveform. The simple experiment shows that this is somehow true. If you fix frequency but change amplitude, you will hear the change of loudness while the pitch stays the same. If you change frequency with a fixed loudness, the most striking is the change in pitch. The switch from A_4 played by piano to A_4 played on violin definitely changes the waveform.

But if you change frequency in a wider range, for example, from 50 Hz to 2000 Hz, with a fixed amplitude, you hear the definite changes in loudness. Table 7.1 relates our subjective concepts to physics parameters.

Table 8.1

Physical parameter	Loudness	Pitch	Timbre	Duration
Amplitude	+++	+	+	+
Frequency	+	+++	+	+
Waveform	+	+	+++	+
Duration	+	+	+	+++

8.5 Loudness and Intensity: Sones

Loudness depends primarily on intensity. The larger pressure variations set the eardrums, the ossicles, and the basilar membrane into vibrations of bigger amplitude. As a result, more hairs of the organ of Corti are bent, causing more signals to be sent and processed in our brain.

The decibel scale, which is the closest to human perception out of all the measurable scales, still gives only a rough approximation to listeners' estimates of their own sensations of loudness. The listener should be exposed to sounds of many different amplitudes to judge directly how much loudness corresponds to each. Usually, the subjects of these experiments should evaluate the relative loudness of two sources (say, source A is twice as loud as source B) and all these experiments should be performed on big groups of people to provide statistical reliability.

As a result, a loudness scale was developed in which the unit of loudness is called a **sone**.

The sone is defined as the loudness of a 1000 Hz tone at a sound level of 40 dB.

Sounds of other loudness are compared to that. Fig. 8.10 shows several dependences of loudness for different frequencies on SIL in decibels. The graph shows that our sensation of loudness depends on frequency: points A, B, and C have the same loudness, but for point A, at 4000 Hz, we should have SIL of 35 dB; for point B, at 1000 Hz, 40 dB; and for point C, at 50 Hz, more than 70 dB.

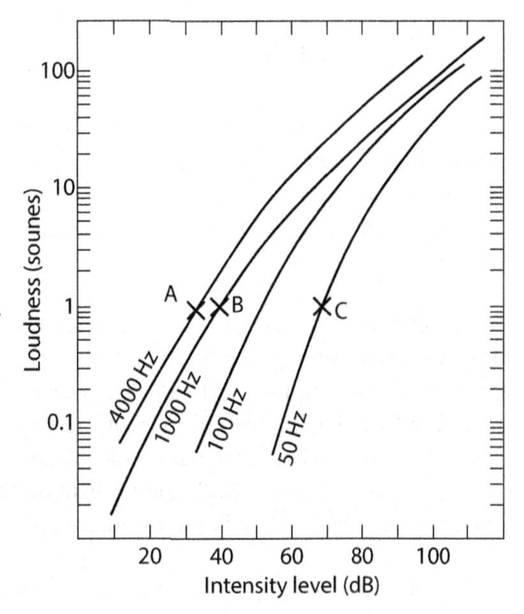

Fig. 8.10. Dependence on loudness in sones on SIL in db. Three points of equal loudness are shown.
Source: Musical Acoustics by Donald Hall, 2002 Cengage Learning, Inc.

For the intensities and frequencies of musical interest (above 50 dB), it is safe to say that the increase of SIL by 10 dB doubles the loudness of sound. The increase of SIL by 10 dB corresponds to multiplying intensity by a factor of 10. This means that the group of 10 performers all singing the same note sound about twice as loud as one performer. And 100 performers sound 4 times as loud as one performer.

8.6 Pitch and Frequency

As we know from Section 8.1, different frequencies set into vibration different parts of the basilar membrane, so our brain receives signals from different sets of neurons, interpreting these signals as different pitch. This is a very simplified picture, but it is good for the beginning.

For the purposes of comparison of different pitches, we cannot use a group of just any people—we need a significant group of musicians with perfect ears. The natural unit for their judgment is the octave, which corresponds in Physics to the doubling of the frequency.

The octave is subdivided into other musical intervals, as we considered in Chapter 7. The number of intervals can be different, but European music usually divides the octave into 12 pitches, called semitones.

The pitch depends on sound level only slightly. Low pitches usually sound a bit lower and high pitches a bit higher at very high sound levels. For instance, a change in pitch of 240 Hz sine wave with the increase of sound level from 40 dB to 90 db was only ¾ of semitone, and for complex not sine wave, even smaller. This is very fortunate for performers and listeners: musical performance would be very difficult with substantial changes of pitch during changes of the sound level.

8.7 Pitch and Loudness: Phons

So, the sound level has relatively small influence on the pitch. The pitch is almost completely defined by frequency. But we already know from Section 8.5 that frequency has a pretty big effect on the sensation of loudness. There are two independent parameters that define loudness: sound level and frequency. We want to understand how loudness depends **simultaneously** on both of them.

The result of such research is a set of graphs with measurable parameters of axis: frequency (in Hz) and sound intensity level (in dB). Fig. 8.11 shows the family of these graphs for perfect sine waves. This diagram is called Fletcher-Munson diagram, and the curves on it represent **the equal-loudness levels**. This means that each curve represents the set of sine waves of different sine waves with such a combination of intensity and frequency that they sound for us equally loud. For example, points A, B, and C all sit on the same curve. This means that 100 Hz 60 dB sine wave A, B, and C have the same loudness.

Each contour in Fig. 8.11 is marked with a corresponding number of a unit that is new to us: the **phon**.

The loudness level in phons is the same number as the sound intensity level for 1000 Hz in dB. This means that for a wave of 1000 Hz if we know the SIL, say 80 dB, we automatically know the loudness in phons: 80 phons. For all other frequencies we should refer to Fig. 8.11.

Example 8.1. We have three sine sound waves, all of SIL 60 dB and frequencies 100 Hz, 1000 Hz, and 3000 Hz. Which of them is the loudest and which is the least loud?

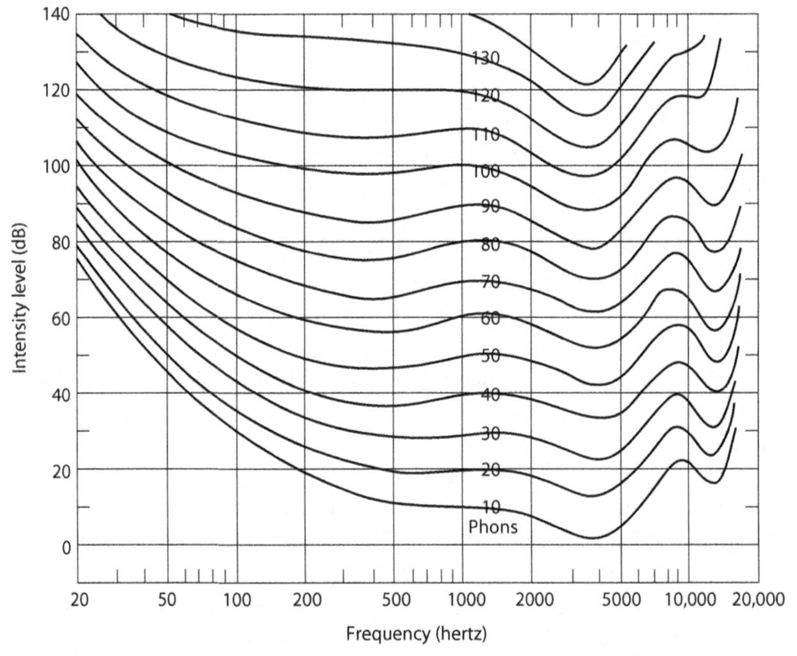

Fig. 8.11. The Fletcher-Munson diagram. The lines of equal loudness are labeled in phons.

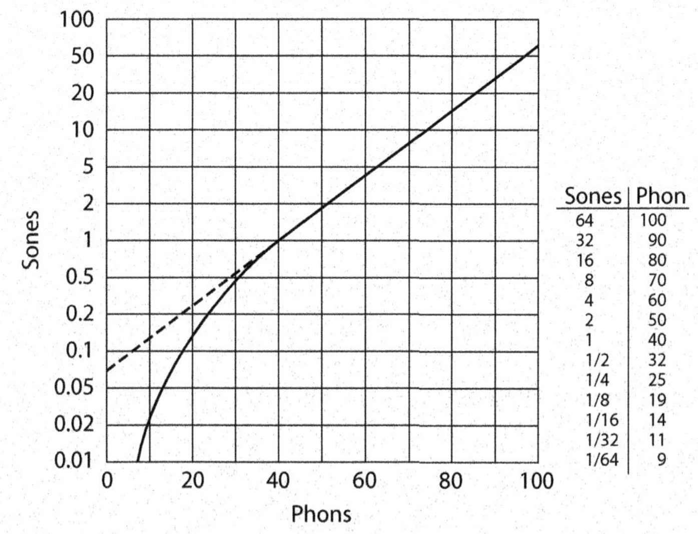

Sones	Phon
64	100
32	90
16	80
8	70
4	60
2	50
1	40
1/2	32
1/4	25
1/8	19
1/16	14
1/32	11
1/64	9

Fig. 8.12. The mutual correspondence of sones and phons.

Answer:

Refer to the diagram in Fig. 8.11. A 60 dB 100 Hz wave has the loudness 50 phons; 1000 Hz, 60 phons; 3000 Hz, between 60 and 70 phons. Hence, the loudest wave is the wave of frequency 3000 Hz, the least loud is the wave of frequency 100 Hz.

Example 8.2. Three sine waves of frequencies 100 Hz, 1000 Hz, 3000 Hz have the same loudness 70 phons. To what SIL in dB does each of them correspond?

Answer:

Refer to the diagram 7.11. SIL for a 100 Hz 70 phons wave is between 70 and 80 dB, approximately 75 dB; SIL for 1000 Hz is 70 dB, SIL for 3000 Hz is slightly bigger than 60 dB.

The graph of the correspondence of sones and phons is shown in Fig. 8.12 together with a table for some specific numbers. Thus, we determine loudness in sones in two steps: first we find the amount of phons from diagram 8.11, then we look for the corresponding amount of sones with graph (table) 8.12.

Example 8.3. What is the loudness in sones for a sine wave of frequency 200 Hz and sound level 80 dB?

Answer:

First refer to the diagram 8.11. The loudness in phons for a wave of frequency 200 Hz, 80 dB is 80 phons. Now check the graph in Fig. 8.12: the loudness 80 phons corresponds to 16 sones.

8.8 Timbre and Waveform

The waveform is the primary factor that defines a timbre. We discuss this in more detail in Chapter 9.

Summary, Terms, and Relations

Human ear, important parts: pinna, ossicles, basilar membrane, organ of Corti.

Just noticeable differences (JND) in frequency and loudness.

Units of loudness:

The sone is defined as the loudness of a 1000 Hz tone at a sound level of 40 dB.

The loudness level in phons is the same number as the sound intensity level for 1000 Hz in dB.

Questions and Exercises

1. Discuss the parts of the outer, middle, and inner ear and their roles in sound perception.
2. At a low enough sound level, the JNDs in sound may become as large as 2 dB. What is the minimum detectable change in intensity?
3. Suppose you change a sine wave frequency from 200 Hz to 450 Hz, and then from 450 Hz to 700 Hz. What change will be perceived as the bigger change in pitch?
4. What loudness level in phons corresponds to a loudness of (a) 2 sones, (b) 4 sones, and (c) 10 sones?

6. If you hear three sounds with frequency 50 Hz, 2000 Hz, and 4000 Hz at the same SIL 60 db, which will sound loudest and which the quietest?

5. If you hear sounds with frequency 50 Hz, 2000 Hz, and 4000 Hz, each at the same loudness level 60 phons, which actually has the largest intensity and which the least?

7. What sound level in dB is required for a 10 sone loudness at frequencies (a) 100 Hz, (b) 1000 Hz, and (c) 3000 Hz?
8. What is the loudness in sones for the following sine waves: (a) 100 Hz at 40 dB, (b) 4000 Hz at 80 dB, and (c) 2000 Hz at 60 dB?
9. For all levels to sound equal in loudness to 40 dB, 1000 Hz, what sound levels in dB would be required at (a) 100 Hz, (b) 300 Hz, and (c) 5000 Hz?

CHAPTER 9

Fourier Analysis of Simplest Sound Spectra

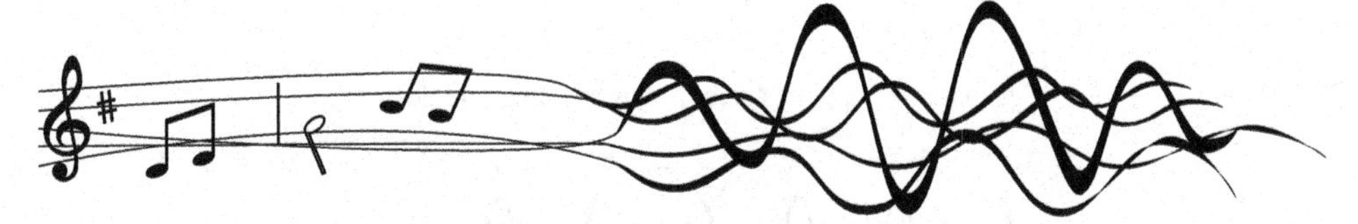

9.1 What Is a Spectrum?

In Latin **spectrum** means "image" or "apparition," including the meaning "spectre" or "ghost." So, what kind of ghost do we have in this application to science?

We already are used to graphs showing the oscillatory process as function of time. It can be pressure variation in a sound wave or displacement of a mass on the string from equilibrium position. But the form of these functions can be sophisticated and unclear. See, for example, Figs. 9.1a, b, and c. Figs. 9.1a and b have the clear periodic behavior, but its form is far from the simple form of sine function, as described in Chapter 3 and shown again in Fig. 9.1c. If the process shown is a sound wave that falls in the audible range of frequencies, we will hear a steady sound in all three situations, but the character, or timbre of sound, will be different.

In Physics, instead of graphs showing the oscillatory process as function of time, we use the graphs to show the frequency content of the function, which means **the signal as a function of the frequency**. See, for example, the simplest spectrum of sine wave (Fig. 9.2). This function contains only one frequency, which is shown with a vertical line. The height of this line corresponds to the amplitude (or intensity, or sound intensity level) of our wave.

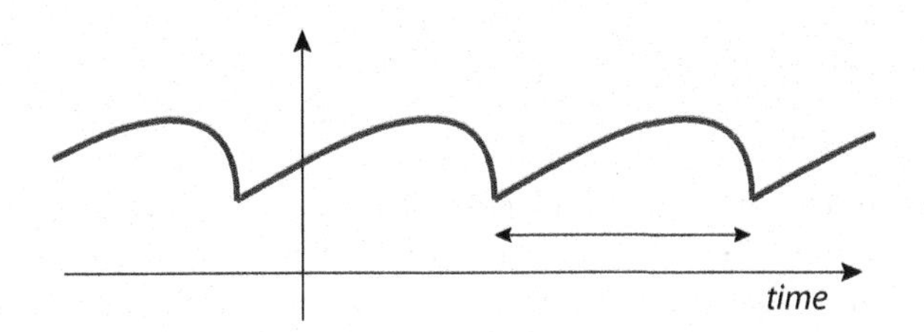

Fig. 9.1a. Example of periodic but not sinusoidal function.

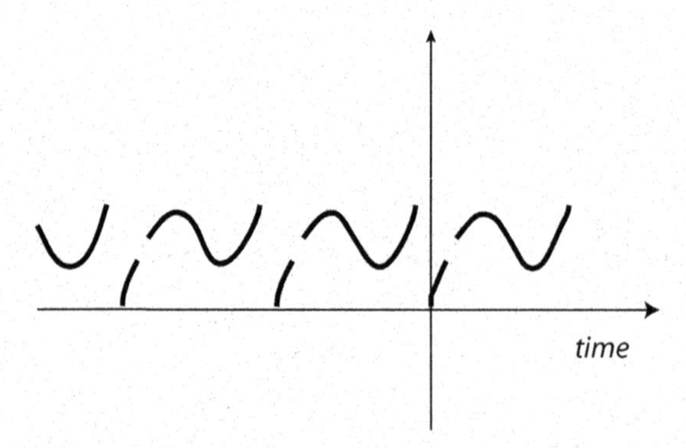

Fig. 9.1b. Example of periodic but not sinusoidal function.

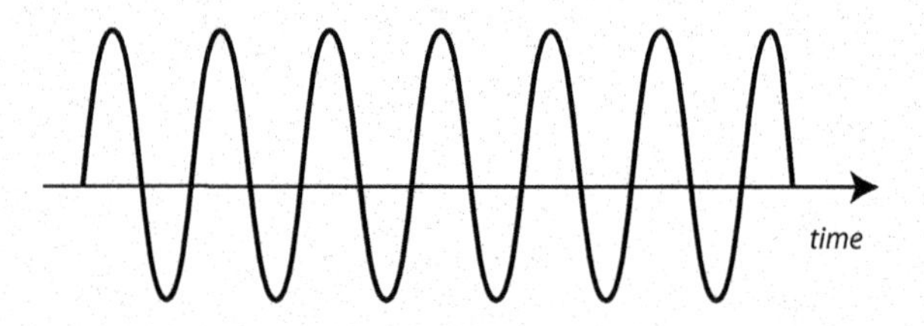

Fig. 9.1c. Sinusoidal function.

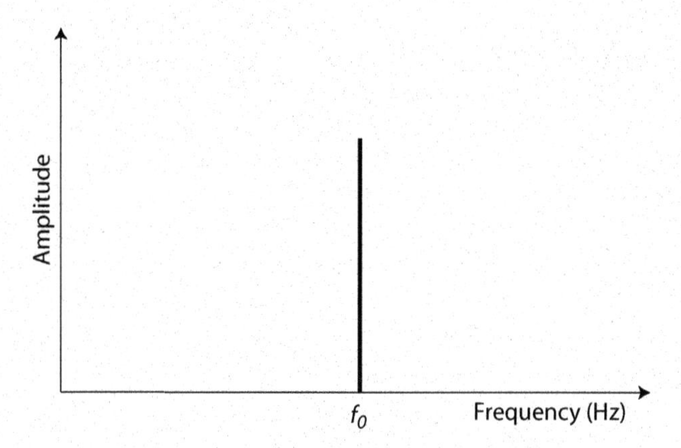

Fig. 9.2. Spectrum of sinusoidal function: the single line corresponding to the frequency of repetition of the sine wave.

Fig. 9.3a shows two sine functions of different amplitude and frequency. For example, two tuning forks played together. The spectrum of this is shown in Fig. 9.3b: it has two vertical lines, one per sine function. The height of these lines corresponds to the sound intensity level of each tuning fork. Note that so far we discussed the spectra of only sine functions. The reasons for this will be explained later.

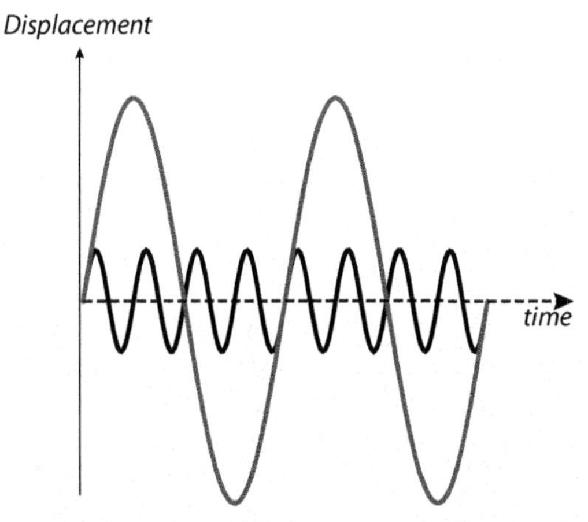

Fig. 9.3a. Two sinusoidal sine waves of different frequencies and amplitudes.

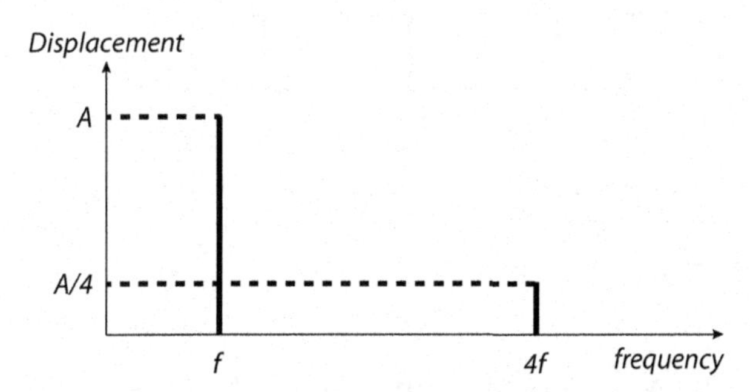

Fig. 9.3b. Spectrum of two sinusoidal functions is shown in Fig 9.3a. Two lines correspond to two different frequencies. The height of these lines corresponds to the amplitude of functions in 9.3a.

The main goal of this chapter is to show and to explain how the seemingly complex waves, shown in Fig. 9.1a and b can be considered and understood in terms of simple sine waves.

9.2 Visualization of Steady Tones

How can we imagine different periodic waves?

The **sine** wave (Fig. 9.1c) is always created when we have some process of simple harmonic motion: the position of a mass on a spring follows time as a sine function as well as the simple pendulum in a grandfather clock. The prongs of a tuning fork perform nearly sinusoidal motion too. The voltage in the outlet in your apartment behaves with time that is also almost sinusoidal.

The **square wave** (Fig. 9.4a) can be made by simply flipping a light switch in your room back and forth. For example: one second light on, one second light off. The illumination as a function of time will have a form shown in Fig. 9.4a. The square wave can be easily generalized in the form of the **pulse wave**. In the pulse wave the time when the light is on is not equal to the time that the light is off.

The **triangle wave** (Fig. 9.4b) could be created following our analogy by gradually increasing and decreasing the illumination in a room. The simple **sawtooth wave**, also a member of a family of triangle waves, is an extreme example, depicting when the illumination is increasing gradually and then the light momentarily switches off (Fig. 9.4c).

Random noise can be visualized by randomly increasing and decreasing the illumination (Fig. 9.4d). This signal differs from all the above-mentioned signals because its waveform is not periodic, which means that it is impossible to find an element that repeats itself. Yet, it is a steady, continuous signal. In terms of sound it will be felt as unorganized, non-pleasant, or just noise.

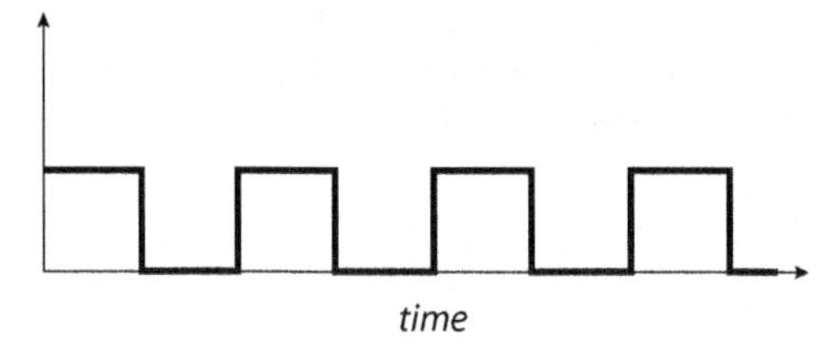

Fig. 9.4a. Square wave.

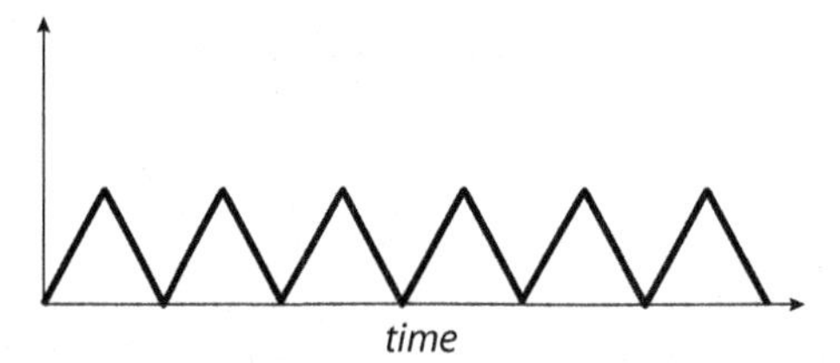

Fig. 9.4b. Triangle wave.

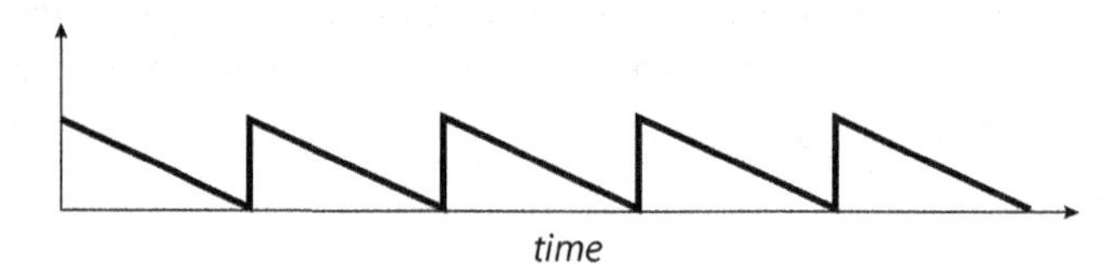

Fig. 9.4c. Sawtooth wave.

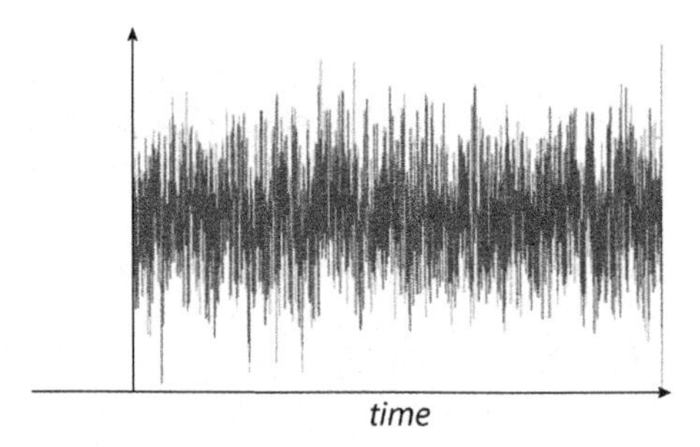

Fig. 9.4d. Random noise.

We want to find a way to describe all these forms of waves in common terms. For this we should answer a simple question: which of all forms, shown in Fig. 9.3 and 9.4, has the simplest possible form of **spectrum**? This means, whose dependence of amplitude on frequency is the most trivial to build on **their combination** of all other forms? From the first glance, the square wave looks very promising: what could be easier than a square? But Mother Nature does not know momentarily increasing and decreasing a signal, which form a square. All the simplest possible oscillating systems, such as a mass on a spring and the motion of a tuning fork's prongs, demonstrate **sinusoidal** character.

9.3 The Spectra of Simplest Periodic Waves

We know that any wave that repeats itself is called a periodic wave; its period T is the length of time it takes to complete the basic pattern and its frequency f = 1/T is how many times per second that whole pattern repeats. In this subchapter we will combine several sinusoidal waves to see how the complex waves will look. This will help us later in a much more complicated task: to break complex periodic waves into simple components.

For starters, we will take two sinusoidal waves of different frequencies and let them run at the same time. The main question here is: is the result also periodic? The answer is, in general, no. If the two frequencies we are trying to mix do not belong to the same harmonic series or if, in other words, the ratio of these frequencies is not close to the ratio of integers, then the combined wave is nonperiodic, as shown in Fig. 9.5a. For this particular figure we took $f_1 = 261.6$ and $f_2 = 370.0$ Hz. The ratio of these frequencies is 1.41437, it is far from integers, and the second wave completes the whole cycle plus another odd fraction each time the first wave completes one whole cycle. If we start them together at time zero, we could wait virtually forever without having both start another cycle at precisely the same time.

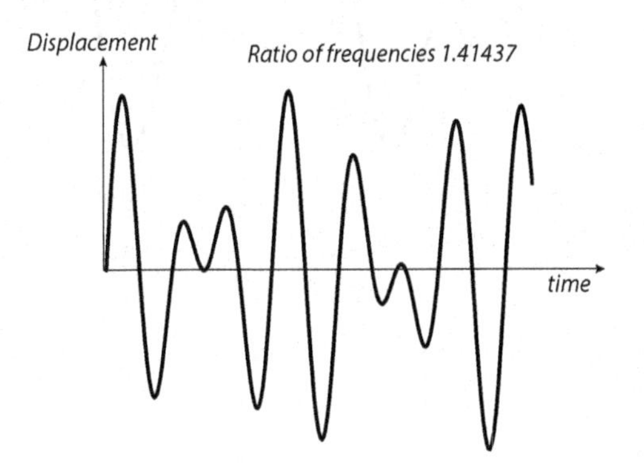

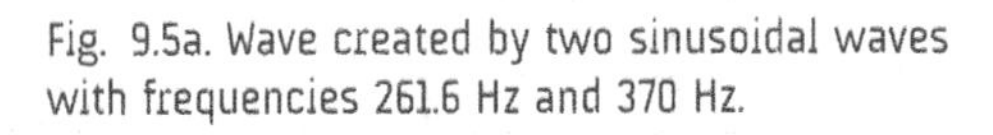
Fig. 9.5a. Wave created by two sinusoidal waves with frequencies 261.6 Hz and 370 Hz.

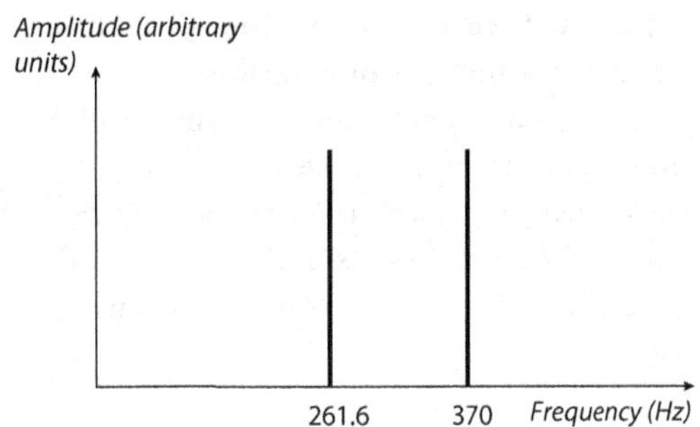

Fig. 9.5b. Spectrum of the combined wave shown in Fig 9.5a.

But if the frequencies of two waves are more carefully chosen, we can get a periodic combination. If, for example, f1 and f2 were 220.0 Hz and 440.0 Hz, every cycle of the first wave (Fig. 9.6a) will take the same time as exactly two cycles of the second wave (Figs. 9.6b and c). It is not necessary that both start from zero at the same time. We can shift one wave forward or backward in time relative to another (Figs. 9.7a, b, and c). As long as we leave the frequencies the same, the resulting complex wave will have a different shape but still the same periodicity.

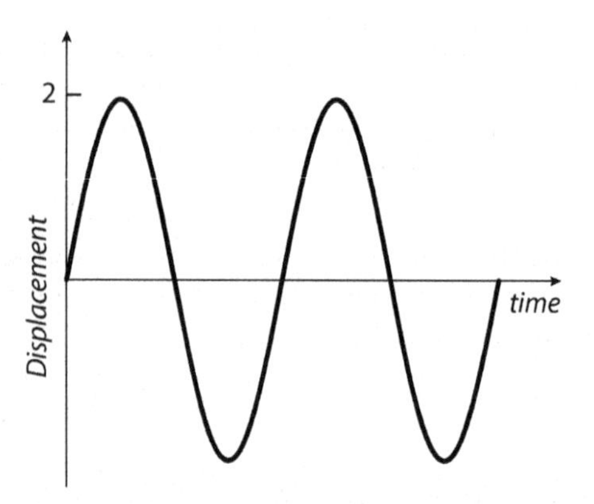

Fig. 9.6a. Sinusoidal wave of frequency 220 Hz.

Fig. 9.8a shows six sinusoidal waves, the frequencies of which belong to the same harmonic series with the fundamental 100 Hz. The result of adding these six members (Fig. 9.8b) is also periodic, with a frequency of repetition equal to the fundamental of our harmonic series. Note that we added waves of 200 Hz, 300 Hz, 400 Hz, 500 Hz, 600 Hz, and 800 Hz. The fundamental 100 Hz is missing, yet the repetition frequency is 100 Hz.

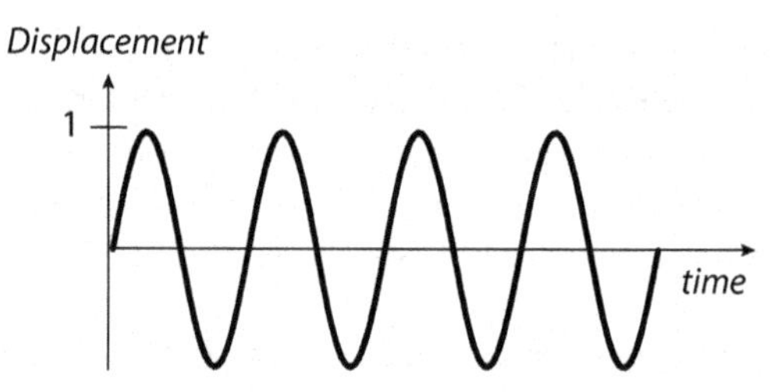

Fig. 9.6b. Sinusoidal wave of frequency 440 H z (twice as big than in Fig. 9.6a and twice as small as than in 9.6a amplitude).

After these two illustrations we can make a general statement about the addition of sinusoidal waves:

A periodic complex wave is one that has the same repetition frequency as the series fundamental. These types of waves are created when a set of sine waves has frequencies that are in a harmonic series. The individual parts can have any amplitudes, and any may be shifted relative to each other on any arbitrary fraction of the period. The amplitudes and relative shift govern the shape of the complex resulting wave.

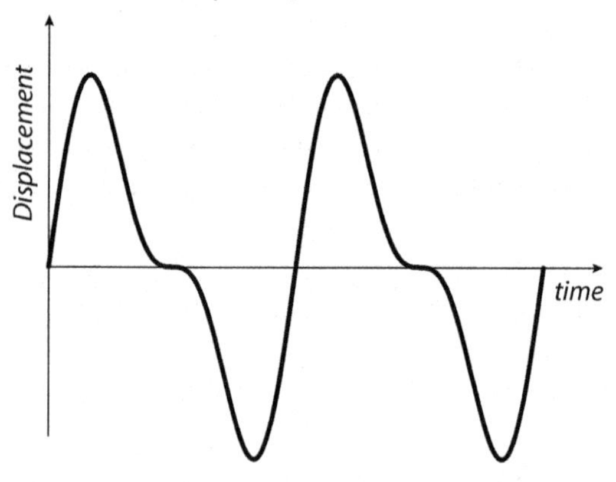

Fig. 9.6c. Combined wave created by waves from Fig 9.6a and 9.6b.

Significantly, we can invert this procedure. We can take any steady periodic wave and break it down into its sinusoidal elements. In order to do this, we must use sinusoids which form a harmonic series.

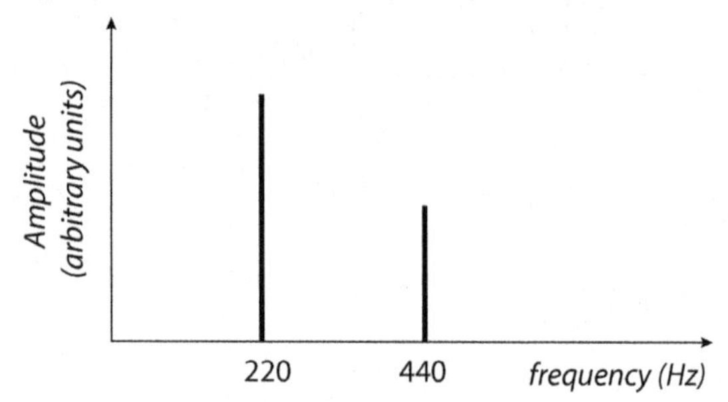

Fig. 9.6d. Spectrum of combined wave shown in Fig. 9.6c.

The theorem that explains this process was developed by the French mathematician Jean Baptiste Joseph Fourier in 1822, and since then everything about spectra comes back sooner or later to his theorem. Combining the sine waves to make complex waves is called the **Fourier synthesis**; the inverse procedure of resolving complex waves into sine components is called **Fourier analysis**. Dependence of amplitudes of corresponding sine components on frequency is called **Fourier spectrum**. Thus, consider:

If we have a set of sine waves whose frequencies form a harmonic series with fundamental 1/T, we can create any periodic wave of period T. The shape of the complex waveform tells us if we have the precise amplitude and relative shift necessary to create a particular waveform.

In a similar fashion, nonperiodic waves also can be represented as sums of Fourier components.

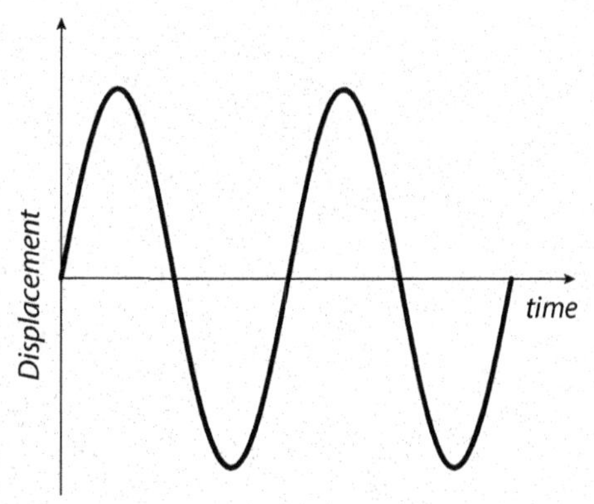

Fig. 9.7a. Sinusoidal wave of some arbitrary frequency.

Fig. 9.7b. Sinusoidal wave of the frequency twice as big as that in Fig. 9.7a. Note that this second wave is shifted relatively to the wave shown in Fig. 9.7a.

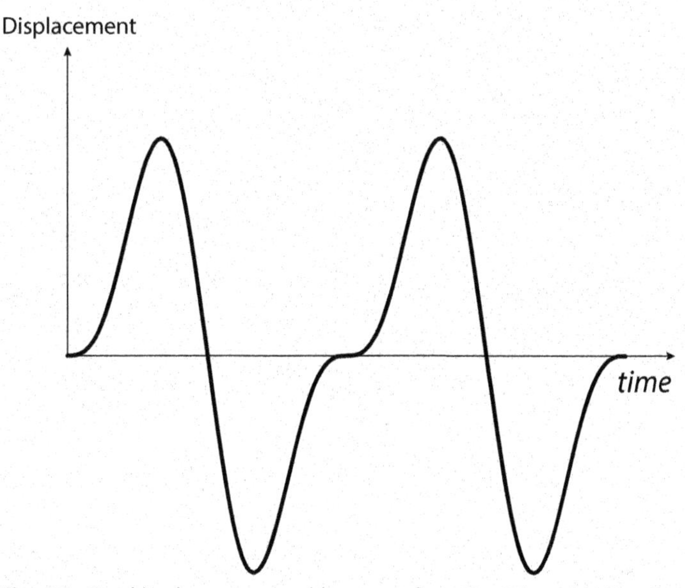

Fig. 9.7c. Combined wave created by waves from Fig. 9.7a and b. Note that combined wave is periodic again like in Fig. 9.6c.

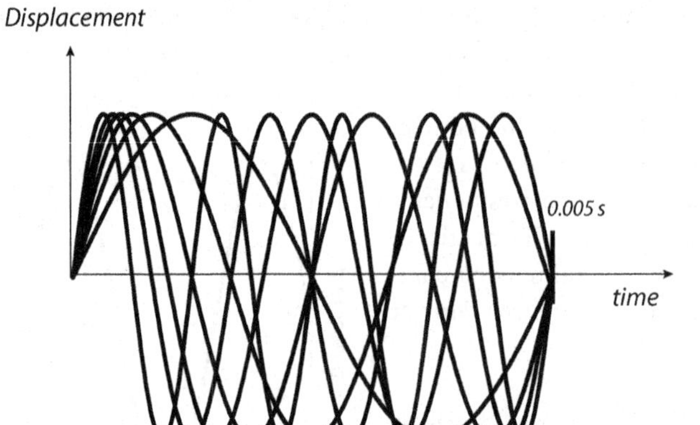

Fig. 9.8a. Six sinusoidal waves belonging to the same harmonic series.

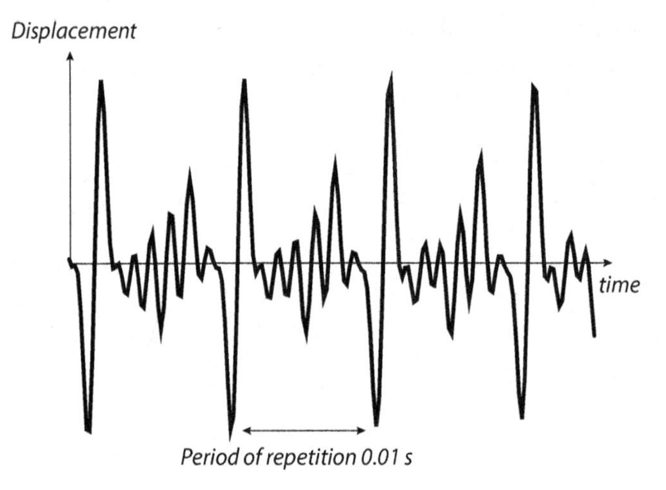

Fig. 9.8b. Resulting wave obtained by addition of sinusoidal waves from Fig. 9.8a. The period of repletion is shown.

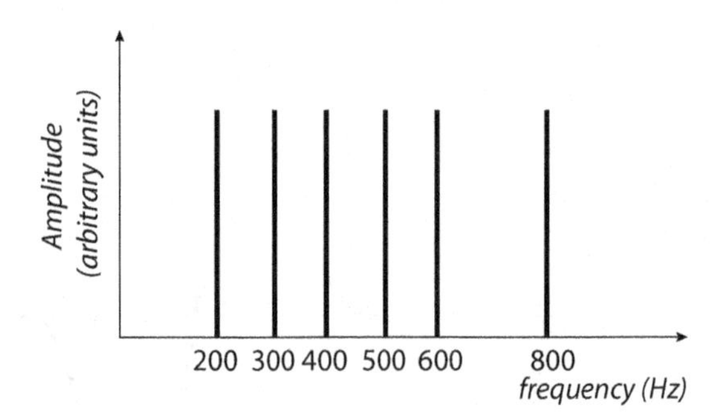

Fig. 9.8c. Spectrum of periodic wave as shown in Fig. 9.8a.

If the considered wave is the sound wave, it will be perceived aurally as "contaminated" or unstable. The inverted procedure for a nonperiodic wave is also applicable:

A nonperiodic, complex wave can be generated by any combination of sine waves whose frequencies do not belong to a harmonic series. Again, we must have the precise amplitude and relative shift necessary to create this particular waveform.

Fig. 9.5b demonstrates the spectrum of a nonperiodic wave. It has two vertical lines of corresponding length, as in Fig. 9.6d, but the frequencies do not belong to a harmonic series.

Let us check the Fourier spectra of waves considered for illustration in this subchapter of the wave shown in Fig. 9.6b. We combined two sine waves. The spectrum of this combination is shown in Fig. 9.6d. It has two vertical lines at position 220 and 440 Hz. The height of these lines is equal to the amplitude of the corresponding components.

Fig. 9.8c shows not two, but six vertical lines because we constructed the wave in Fig. 9.8b out of six components. Again, the height of each vertical line corresponds to the amplitude of the Fourier components.

Example 9.1. Three sinusoidal waves of frequencies 150, 180, and 270 Hz are combined. Is the resulting wave periodic or nonperiodic? If the wave is periodic, what is the frequency of repetition?

Answer:

This wave is periodic, with frequency of repetition equal to biggest common divisor of numbers 150, 180, and 270: 30 Hz.

Example 9.2. Sketch a wave whose spectrum is shown in Fig. 9.6d. What frequency of repetition does this wave have?

Answer:

This is a periodic wave of complex form, with frequency of repetition equal to 220 Hz.

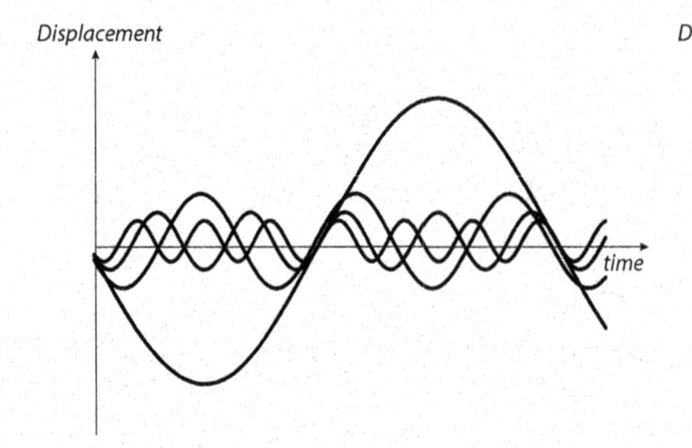

Fig. 9.9a. Four first harmonics to form a square wave. Note that only odd harmonics participate in the creation of square form.

Fig. 9.9b. Result of addition of four first harmonics as shown in Fig. 9.9a. The square form already can be recognized.

9.4 The Spectrum of the Square Wave

The form of the square wave is shown in Fig. 9.4a. Now we should understand what the recipe is for combining sine waves to produce a wave of such form. Because we do not use advanced Mathematics in this book, we illustrate it with Figs. 9.9.a and b. In part (a) this figure shows the sinusoidal waves of frequencies 100 Hz, 300 Hz, 500 Hz, 700 Hz. Note that even multiples of 100 are missing and that amplitude of each next component (or harmonic) is gradually decreasing. The result of the addition of these 4 components is shown in part (b). As we may already see, the result looks pretty much like a square wave. If we add sine waves of frequencies 900 and 1100 Hz, this square form will become even more obvious.

So, the square wave can be created by adding odd-numbered components (harmonics). The spectrum of a square wave is shown in Fig. 9.10. The amplitudes of components decrease as 1/n, where n is the number of a harmonic, and the intensities of the components decrease as $1/n^2$. The SIL in the square wave falls by 6 dB per octave.

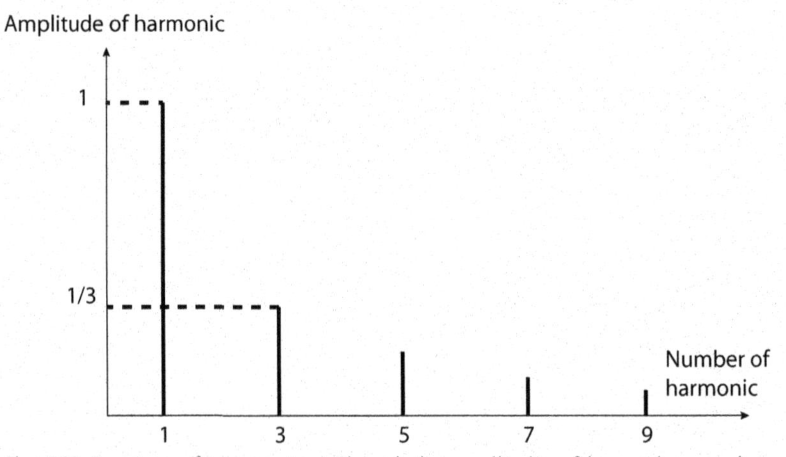

Fig. 9.10. Spectrum of square wave. The relative amplitudes of harmonics are shown.

Example 9.3. The amplitude of a first harmonic in a square wave is 100 units. What is the amplitude of the fifth harmonic? Of the fourth harmonic?

Answer:

Fifth harmonic of the square wave is equal to 100 units/5 = 20 units. The square wave does not have even harmonics, so the amplitude of the fourth one is zero.

Example 9.4. The intensity of a first harmonic of a square wave is 10 W/m^2. What is the intensity of the third harmonic of this wave? Of the second one?

Answer:

The intensity of the third harmonic of the square wave is 10 $W/m^2/3^2$ = 1.1 W/m^2. The square wave does not have even harmonics, so the intensity of the second one is zero.

9.5 The Spectrum of the Triangle Wave.

The triangle wave is shown in Fig. 9.4b. This wave by its form is much closer to a sinusoidal wave, so we may expect that components in its spectrum will decrease with the number of harmonics much faster than for a square wave. The spectrum of a triangle wave is shown in Fig. 9.11. It also contains only odd-numbered harmonics. The amplitudes of harmonics decrease as $1/n^2$ now, so the third harmonic will have amplitude not three times less than the first one, as for the square wave, but 3 x 3 = 9 times less. The intensities of upper harmonics for a triangle wave are decreasing, with the number of harmonics as $1/n^4$. As a result, the SIL for the triangle wave is falling at the rate of 12 dB per octave, twice as fast as for the square wave.

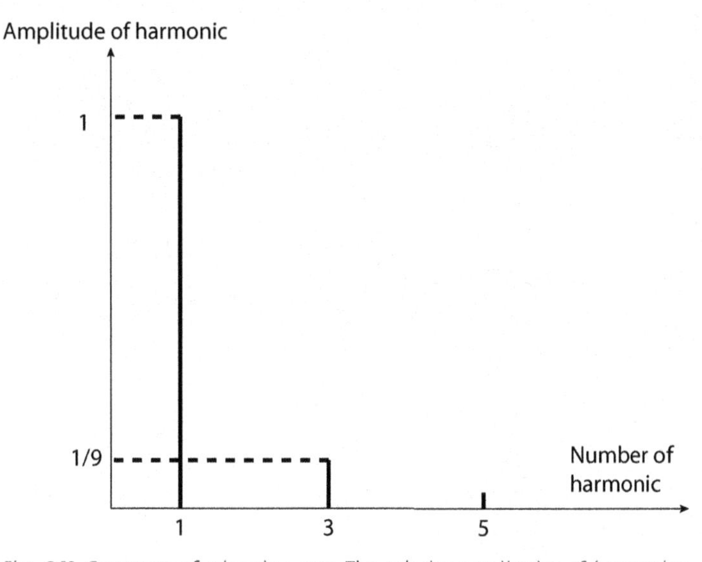

Fig. 9.11. Spectrum of triangle wave. The relative amplitudes of harmonics are shown.

Example 9.5. The amplitude of the first harmonic of a triangle wave is 100 units. What is the amplitude of the fifth harmonic? Of the second one?

Answer:

The amplitude of the fifth harmonic of the triangle wave is 100 units/5^2 = 4 units. The triangle wave does not have even harmonics, so the amplitude of the second harmonic is zero.

Example 9.6. The intensity of the first harmonic of a triangle wave is 10 W/m^2. What is the intensity of the third harmonic? Of the fourth one?

Answer:

The intensity of the third harmonic of the triangle wave is 10 W/m^2/3^4= 10/81 W/m^2.

9.6 The Spectrum of the Simple Sawtooth Wave

The simple sawtooth wave is shown in Fig. 9.5c. This wave has in its spectrum both even- and odd-numbered upper harmonics, which make its sound fuller and richer in comparison to the square wave and the triangle wave (Fig. 9.12). The amplitudes of upper harmonics decrease with a number as $1/n$, and intensities as $1/n^2$, the same as for the square wave. But a square wave does not have even harmonics, and it sounds, as a result, more "flat." The decrease of the SIL level for the simple sawtooth wave is 6 dB per octave.

Example 9.7. The amplitude of the first harmonic in the sawtooth wave is 10 units. What is the amplitude of the second one? Of the fifth one?

Answer:

The amplitude of the second harmonic of the sawtooth wave is 10 units/2 = 5 units. The amplitude of the fifth harmonic is 10 units/5 = 2 units.

Example 9.8. The intensity of the first harmonic in the sawtooth wave is 100 W/m^2. What is the intensity of the fifth harmonic? Of the second one?

Answer:

The intensity of the fifth harmonic of the sawtooth wave is 100 W/m^2/5^2 = 4 W/m^2. The intensity of the second one is 100 W/m^2/2^2 = 25 W/m^2.

Example 9.9. The amplitude of the second harmonic in the sawtooth wave is 20 units. What is the amplitude of the fifth one?

Answer:

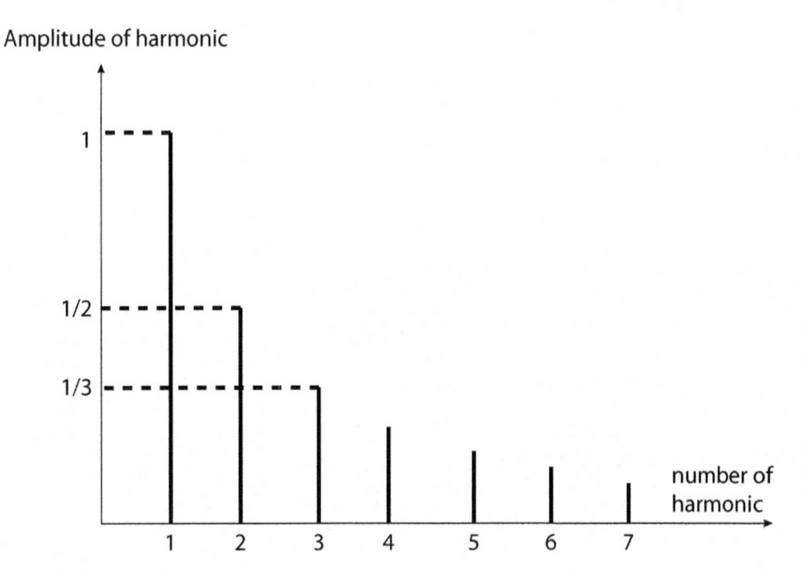

Fig. 9.12. Spectrum of sawtooth wave. The relative amplitudes of harmonics are shown.

First we should define the amplitude of the fundamental: 20 units x 2 = 40 units. The amplitude of the fifth harmonic is 40 units/5 = 8 units.

9.7 The Spectrum of Noise

Random noise, being nonperiodic (Fig. 9.4d), cannot be represented by a harmonic series; instead it requires the combination of sine waves of **all** frequencies.

There are different kinds of random noise. **White** noise is defined as noise whose spectrum includes equal amplitudes of all frequencies (Fig. 9.13a). This is the noise of "static" in your cell phone or the noise between stations on your radio. White noise's hissing sound contains strong high-frequency components, and, as a result, white noise is tiring to listen to. White noise is called so in an analogy with optics. It is well known that white light, being resolved into a rainbow, contains all colors of equal amplitudes.

Pink noise, whose spectrum is shown in Fig. 9.13b, favors lower frequencies, dropping by 3 dB/octave as frequency increases. Pink noise occurs in many physical, biological, and even economic systems. In Physics it is present in some meteorological and seismological data series, the electromagnetic radiation of some astronomical bodies, and in almost all electronic devices. In biological systems, it is present in heart beat rhythms and neural activity. Recently the terms of pink noise were successfully applied to the modeling of problems in mental psychology.

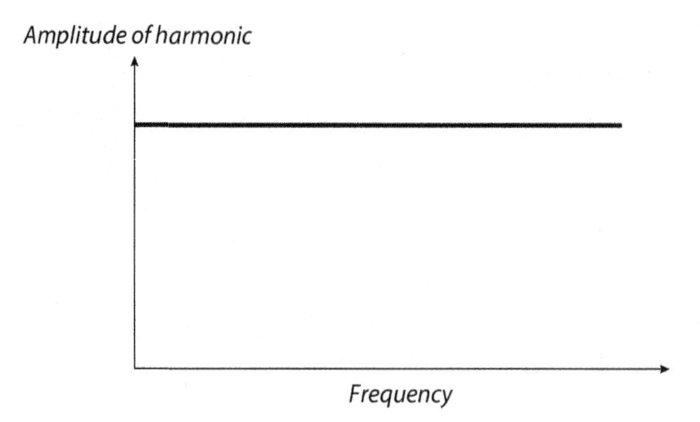

Fig. 9.13a. Spectrum of "white" noise.

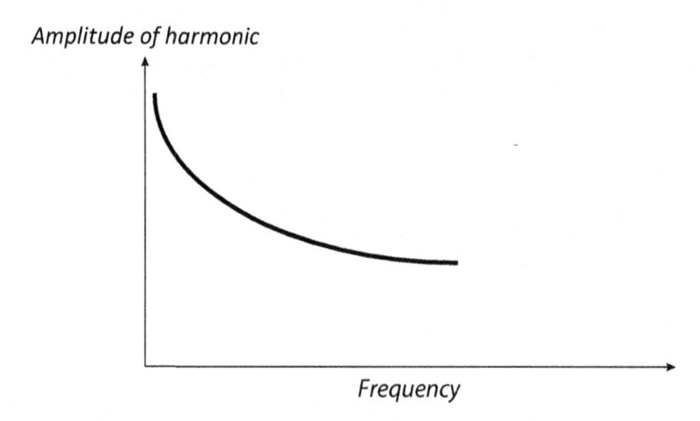

Fig. 9.13b. Spectrum of "pink" noise, favoring low frequencies.

9.8 Modulated Tones

The tones of musical instruments are not perfectly steady. The transient components of sound as well as modulation of the steady tones may create additional features in the spectra.

The simplest kind of unsteadiness in a sound is a change in amplitude while the shape of a wave, as well as frequency of a wave, stay the same. This is called amplitude modulation (AM). If the modulating process is repetitive, the modulated tone still has some sort of steadiness. Fig. 9.14a shows a "carrier" sine wave of high frequency (200 Hz) and substantial amplitude. The next graph (Fig. 9.14b) also shows a sine wave but of smaller amplitude and much smaller frequency (5 Hz). The resulting wave demonstrates periodic changes in amplitude (Fig. 9.14c) with the frequency of the modulating wave at 5 Hz. The spectrum of this simplest modulated wave is shown in Fig. 9.14d. As we see, this spectrum has two lines: the short one, at low frequency, corresponds to a modulation; meanwhile, the strong high-frequency line appears because of the carrier signal.

You may see the abbreviated AM on your car radio. The AM approach is used for transmitting a useful signal (music, news) wirelessly. Commercial AM radio uses strong carrier frequencies of approximately 1 mHz, modulated by a microphone signal at audio frequencies up to 5 kHz. Your radio receiver is tuned to the carrier wave in the mHz range; it reads the modulation and

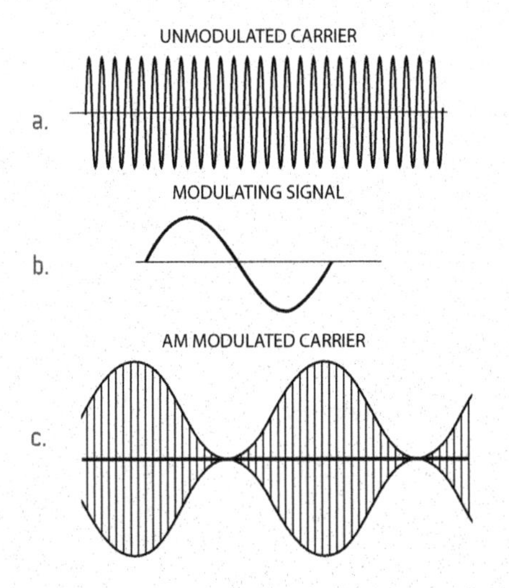

Figs. 9.14.a, b, and c. (a) The "carrier" sinusoidal wave of high frequency. (b) Modulating sinusoidal wave of much smaller frequency than in Fig. 9.14a. (c) Resulting amplitude modulated signal: tremolo.

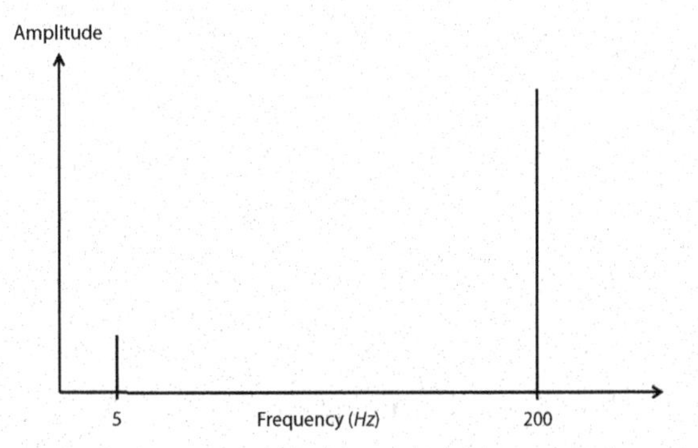

Fig. 9.14d.

uses it to reproduce a sound signal picked up by the microphone in the broadcasting studio.

Steady sound waves of audible frequency also can be carriers; the amplitude modulation of subaudio frequencies, between 1 Hz and 5 Hz, produces the effect of **tremolo**, in which the loudness periodically fluctuates with no change in pitch.

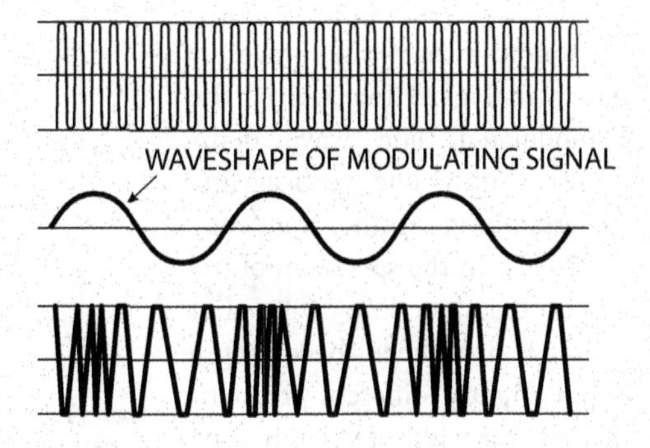

Fig. 9.16. Creation of vibrato: the frequency is periodically changing, meanwhile, the amplitude stays the same. Resulting frequency modulated signal (vibrato) is shown in the lowest graph.

Another kind of repetitive change of a carrier signal is frequency modulation (FM). The frequency of such signal changes periodically. The period of this change is much bigger than the period of a carrier wave. As a result, the total signal looks like the one shown in Fig. 9.15: the wave looks slightly nonperiodic, with repeating compressions and rarefactions.

Commercial FM radio uses carrier frequencies of approximately 100 MHz, which is modulated with audio frequencies up to 15 KHz. This is just another way to transmit the information initially carried by the sound wave by means of the much faster electromagnetic (radio) wave.

The steady sound of audible frequencies also can act as carriers; modulation of their frequencies with subaudio modulating frequencies (between 1 Hz and 10 Hz) produces the effect called **vibrato**. This is characterized by a rate (the modulating frequency, the rate at which you notice the pitch changing) and a strength (the range in which the carrier frequency is altered, which defines how far up and down the pitch changes). Violinists typically use a vibrato rate of approximately 5 to 7 Hz with a variation of approximately 0.2 semitones from average pitch. Note that vibrato created on guitars and violins appears from different physical reasons. On the violin, the performer is changing the effective length of a string, moving a finger back and forth. On the guitar, the effective length of a string cannot be changed because of frets, but the performer is changing the force of tension.

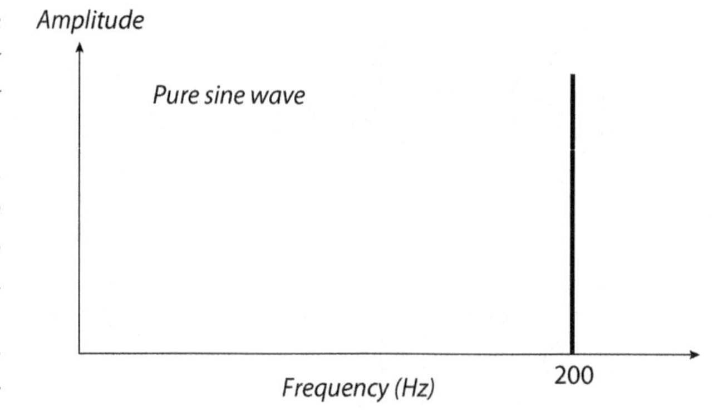

Fig. 9.16a. Spectrum of unmodulated sinusoidal wave.

The spectrum of vibrato in comparison to simple sine waves is shown in Figs. 9.16a and b. The unmodulated sine wave demonstrates only one vertical line corresponding to the frequency of 220 Hz. In the spectrum of the vibrato the same line centered at 220Hz is broadened, becoming a set of slightly different frequencies, with a width determined by the modulated frequency, in the given example of 5 Hz.

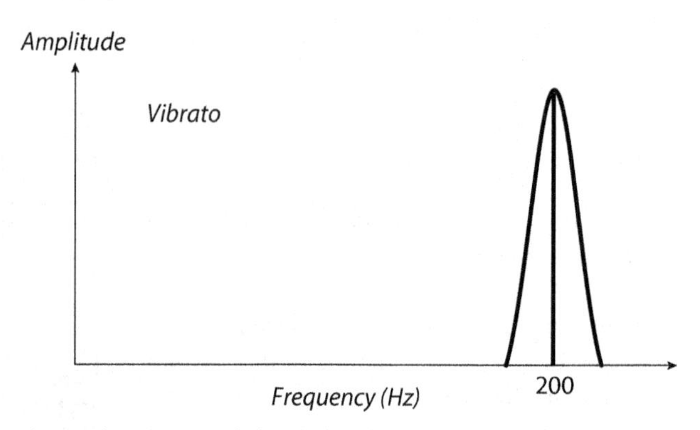

Fig. 9.16b.Spectrum of the sinusoidal wave (sound) performed with vibrato: the line from 9.16a is now broadened.

Summary, Terms, and Relations

Any set of sine waves whose frequencies belong to a harmonic series being added create a periodic complex wave, whose repetition frequency is that of the series fundamental.

Any combination of sine waves with frequencies that do not belong to a harmonic series creates a complex wave that is nonperiodic.

Forms of waves:

A sinusoidal wave has only one harmonic in its spectrum.

Square and triangle waves have only odd harmonics.

A sawtooth wave has both odd and even harmonics.

Tremolo is the amplitude-modulated signal, in which loudness periodically changes with time.

Vibrato is the frequency modulated signal, in which frequency periodically changes with time.

Questions and Exercises

1. If sine waves of frequency 60, 150, 180, and 300 Hz are mixed together, what is the frequency of repletion and period of the resulting complex wave?

2. A complex waveform repeats each 2 ms. What are the frequencies of its Fourier components?

3. The note A_2 has a complex waveform. What are the frequencies of its Fourier components? Without looking at Fig. 5 of Appendix B, what is the name of the notes corresponding to the second and fourth harmonics of its spectrum?

4. The amplitude of the first harmonic of the square wave is 14 units. What is the amplitude of the fourth harmonic? The amplitude of the seventh harmonic?

6. The intensity of the first harmonic of the square wave is 1.25 W/m^2. What is the intensity of its second harmonic? Of its fifth harmonic?

5. The amplitude of the third harmonic of the triangle wave is 25 units. What is the amplitude of its fifth harmonic?

7. The amplitude of the first harmonic of the sawtooth wave is 20 units. What is the amplitude of its fourth harmonic? Of its fifth harmonic?

8. The intensity of the first harmonic of the triangle wave is 0.81 W/m^2. What is the intensity of its third harmonic?

9. The range of audible frequencies is between 17 Hz and 17 KHz. The wave A is a sine wave of frequency 5 Hz, the wave B is a sawtooth wave of frequency 5 Hz. Which of these waves (if either) is audible and why?

10. The sine wave A is of the frequency 20 KHz, and wave B—the sawtooth wave—is also of the frequency 20 KHz. Which of this waves is audible (if either) and why?

11. Discuss the characteristic properties of the spectrum of vibrato and tremolo.

Chapter 10

Basics of Percussion Instruments and Normal Modes

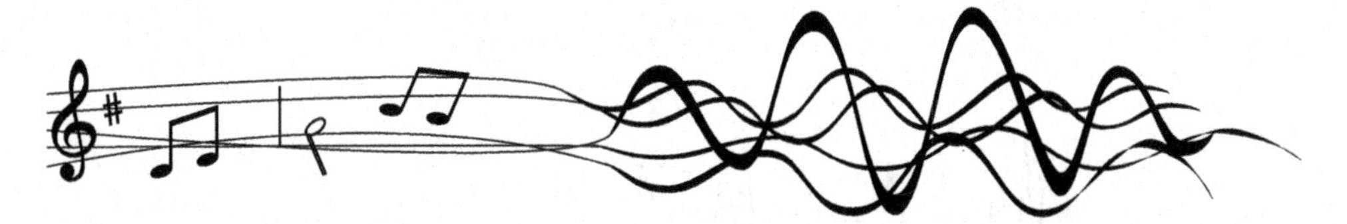

10.1 Discussion of a Problem

We start our study of some particular instruments with the family of percussion and not because they look simpler for us than others. There is no such thing as a "trivial musical instrument." Percussion instruments have very complex spectra, which may or may not contain harmonic series. Percussion instruments are also the only family that widely uses transient sounds. All this makes percussion important to discuss first because we will take some of its general features into the next chapters, for instance, when we discuss the difficulties and special properties of the spectrum of a piano.

We already know that a wave pattern can be represented as a sum of simple sinusoidal harmonics. If the wave is periodic, the sinusoidal waves belong to one harmonic series; if the original

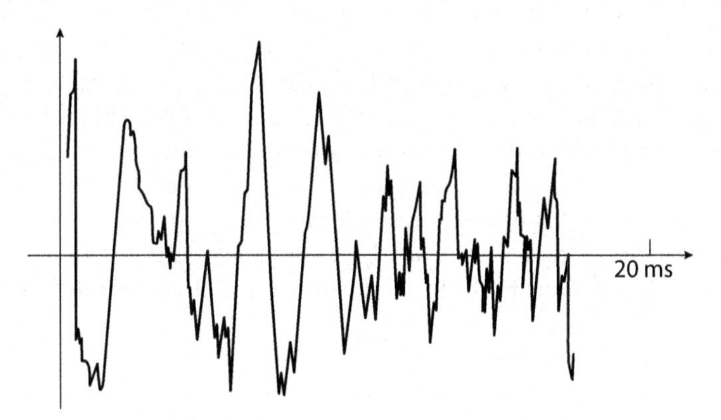

Fig. 10.1a. Oscillations of a point on a big drumhead with time.
Adapted from Musical Acoustics by Donald Hall, 2002 Cengage Learning, Inc.

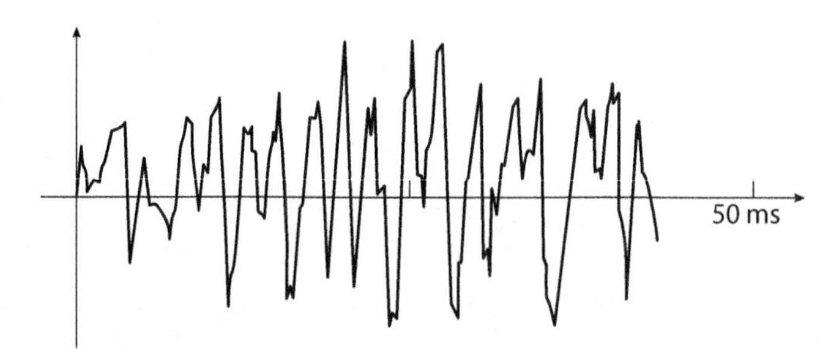

Fig. 10.1b. Oscillations of the tympani with time.
Adapted from Musical Acoustics by Donald Hall, 2002 Cengage Learning, Inc.

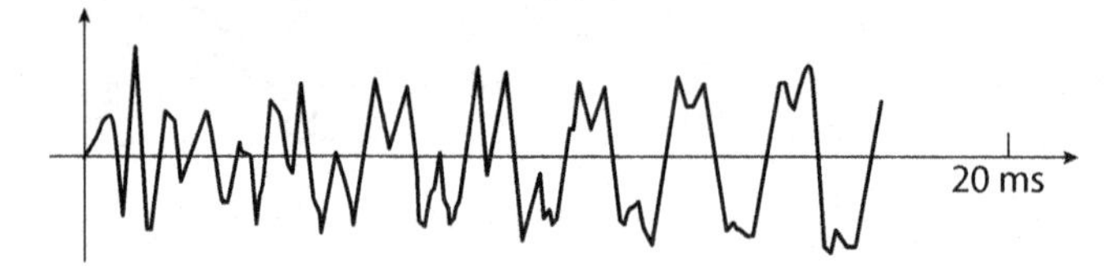

Fig. 10.1c. Oscillations of the xylophone bar with time.
Adapted from Musical Acoustics by Donald Hall, 2002 Cengage Learning, Inc.

wave is not periodic, then we should add simple sinusoidal waves that do not belong to a harmonic series. So at first, the theory looks simple and straightforward. But let us check Fig. 10.1, which shows the oscillations with time of some arbitrary point on (a) the big drumhead, (b) the tympani, and (c) the xylophone. These pictures do not even look periodic to our eye. They are extremely complicated, and it is not exactly obvious how we should approach an analysis.

It is logical to start our approach with the instrument that produces at least a steady musical pitch because we may expect that at least this problem is easier for our consideration. The percussion that is created and used specifically for holding a precise pitch is the tuning fork. But even here we need simplification. The vibration of the tuning fork, as we see from Fig. 10.2a, starts with some irregularities. They are not strong irregularities, as shown in Fig. 10.1, but they exist. The dependence of oscillation of a point on one of the prongs of a tuning fork looks a little bit "hairy." Using material from a previous chapter, we may say that in the spectrum of this device, upper harmonics are present. But if we wait for 5–7 seconds, the oscillations become more and more regular (the amplitude of them is also decreasing, but so far we will pay attention only to periodicity) and after 7 seconds already look almost sinusoidal (Fig. 10.2b). You can easily detect the unsteadiness of a tuning fork sound with your ear. It appears that the features, which make initial oscillations so complicated, are dying out pretty quickly, leaving steady, smooth, and pretty flat sounds of definite pitch. This is the **fundamental tone** of our fork.

So, if we wait 5–7 seconds, the prongs of a tuning fork start to vibrate in simple harmonic motion (SHM) producing sinusoidal waves. But initial unsteady sound is also part of the tuning fork spectrum. It has high and definite pitch. This is called a **clang** tone, and you could even focus on it by mistake if you do not hold a tuning fork close to your ear long enough. We want to represent all the waves and processes based on sinusoidal waves. So the important question here is: maybe this clang tone is also a sinusoidal vibration and what we hear right after the tuning fork was struck is just the mixture of two sine waves? The complication of initial motion of a prong then has a very simple explanation: there are two independent sinusoidal motions (not necessarily with frequencies belonging to the same harmonic series) that occur simultaneously. But if we can prove that

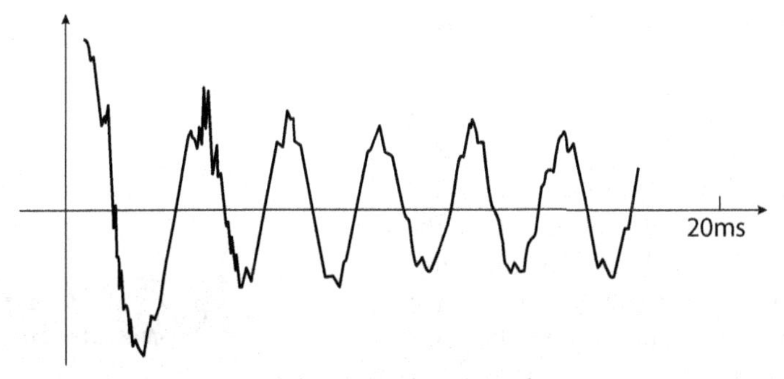

Fig. 10.2a. Oscillations of the tuning from just after excitation. Note that the waveform is far from sinusoidal.
Adapted from Musical Acoustics by Donald Hall, 2002 Cengage Learning, Inc.

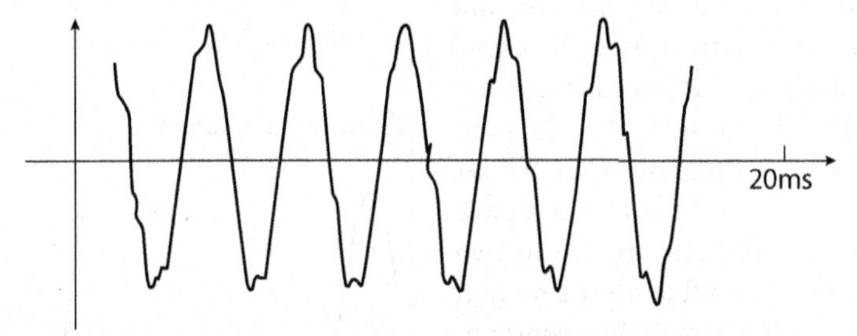

Fig. 10.2b. Oscillations of the tuning fork prongs one second after excitation. The waveform becomes more sinusoidal.
Adapted from Musical Acoustics by Donald Hall, 2002 Cengage Learning, Inc.

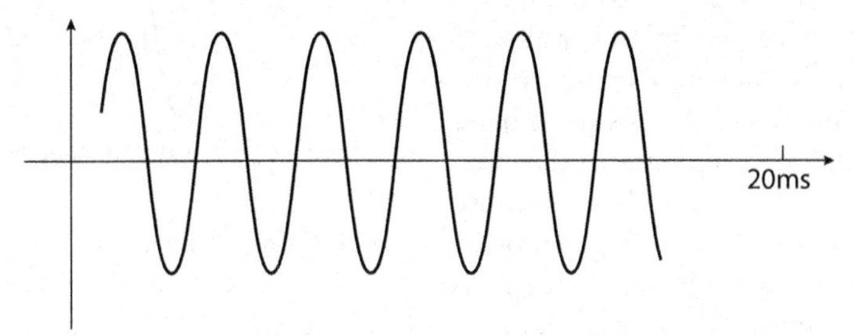

Fig. 10.2c. Oscillations of the tuning fork prongs several seconds after excitation. The waveform is already almost sinusoidal..
Adapted from Musical Acoustics by Donald Hall, 2002 Cengage Learning, Inc.

the initial sound of a tuning fork is created by two simple motions mixed together, then other, more complicated motions, as in Fig. 10.1, might also be created by two or more simple harmonic motions.

Such a hypothesis looks extremely attractive, especially because of its simplicity. But the question pops up: **from where do all these simple motions come?**

To answer this question, we should consider a system so extremely simplified that from the first glance it does not have anything to do with percussion instruments. But don't worry, our idealized system has a very straightforward connection to a tuning fork and other percussion instruments.

10.2 System of Coupled Pendulums

Let us consider a pair of simple pendulums connected with a string (Fig. 10.3a). Without this spring attached, each pendulum swings independently. They have the same length, so the frequency of oscillation also will be the same. Thus, without the spring, the description of a motion of a system is quite simple: the same pattern repeats itself.

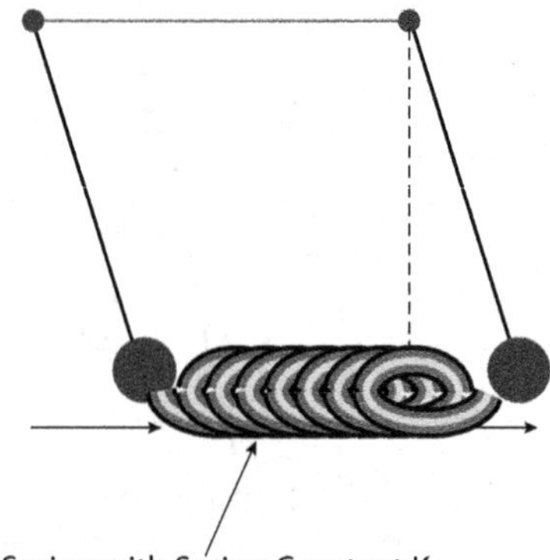

Fig. 10.3a. System of two pendulums connected with a spring of spring constant k.

Now the spring is attached. If we set one pendulum into motion, it will also set into motion the one through the spring. Now each pendulum demonstrates an irregular character of motion (Figs. 10.3b and c). It is not even periodic! Why does it happen? When we set one pendulum aside, we involve, through a spring, also the one, and there are already two different properties that influence the motion of the system: gravity and spring constant. And why should these two know anything about each other?

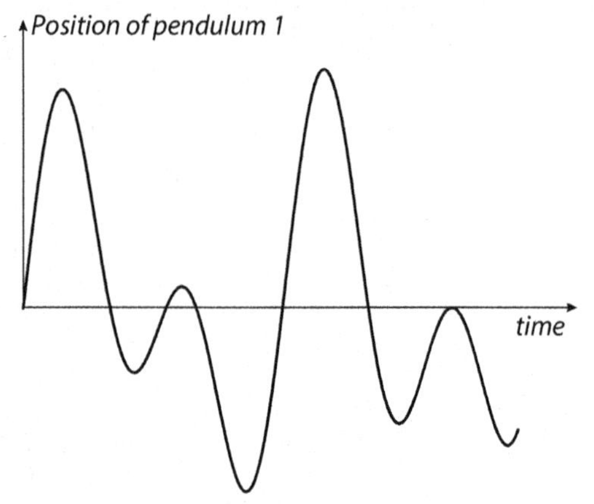

Fig. 10.3b. Motion of first pendulum as a function of time.

Now, let us analyze different possible initial conditions, which eventually will lead to a simple and **periodic** motion. We can perform a lot of experiments, but the answer is quite obvious: we pull both pendulums at the same initial distance to the same side and release them simultaneously. The resulting motion in this situation is exceedingly simple (Fig. 10. 4a): the two pendulums stay in step with each other, the spring is neither compressed nor stretched, so we have SHM with a frequency f_1 equal to the natural one of the lone pendulum.

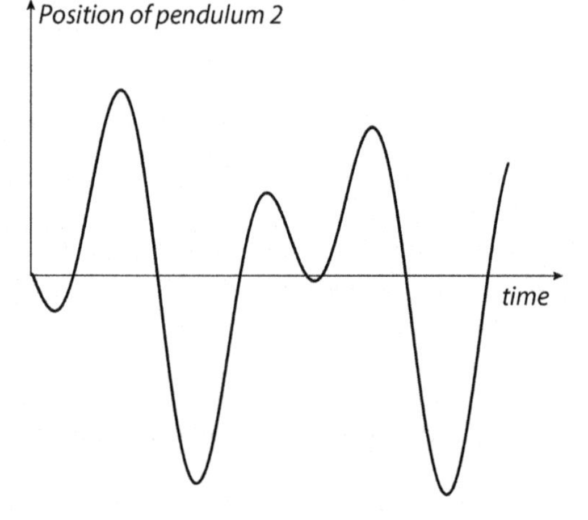

Fig. 10.3c. Motion of second pendulum as a function of time.

But this is not the only way to perform periodic motion. We can start the motion from initial conditions, when two pendulums are pulled at the same distance but to opposite sides, stretching or compressing an attached spring (Fig. 10.4b). In this situation we again have a simple periodic motion, pure SHM, but with another frequency than in the previous example. Now the spring is already in play: it adds additional restoring force, pulling the pendulums back and forth faster than in Fig. 10.4a. So, the frequency f_2 for this second type of motion must be greater than f_1.

The pattern of motion that gives SHM is called normal mode (also natural mode), and each pattern has its own normal mode (natural mode) frequency. The normal modes are properties of the whole system. The frequencies of the normal mode do not necessarily belong to the same harmonic series. Please note that the lowest mode by frequency, even in this simplest of examples, corresponds to all parts of the system (pendulum) moving "in step" with each other.

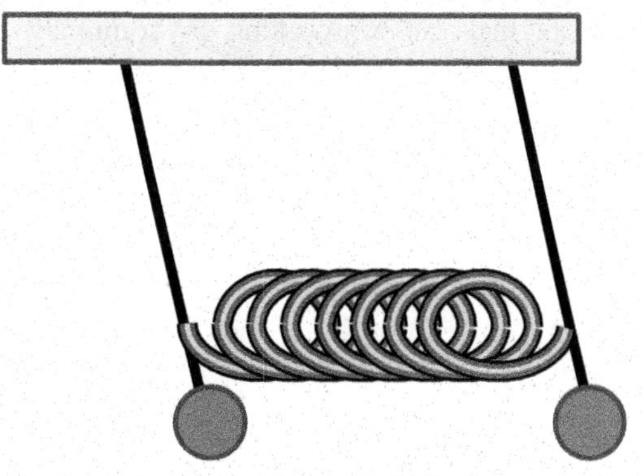

Fig. 10.4a. First normal (natural) mode of two-pendulum system: both pendulums move in phase.

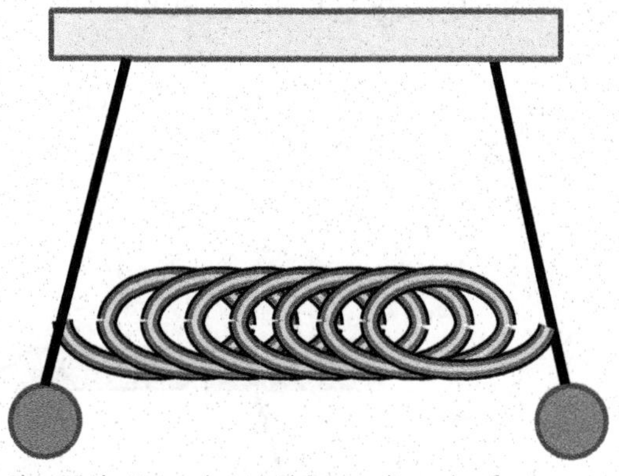

Fig. 10.4b. Second normal (natural) mode of two-pendulum system: two pendulums move out of phase.

Another important conclusion we can make from the example with two pendulums: no matter how hard we try, no matter what kind of initial conditions in which we start, we cannot find the third possible normal mode, demonstrating SHM. So, note that **two** pendulums give us **two** normal modes. Also note, that the excitation of the second mode involves more energy: energy is not stored only in the form of gravitational energy of the pulled-aside pendulums but also in the stretched or compressed spring.

Now let us make the next step in our hypothesis: if the two normal modes represent all possibilities of SHM for the example considered, then maybe we are able to understand **any** motion of our system as the combination of these two.

For example, let us consider a situation when both modes coexist with the same amplitudes. Figs. 10.5 and 10.6 show how we can add them together to find out what motion each pendulum performs individually. The result shows exactly what we had in Figs. 10.3b and c.

It seems that our hypothesis is right, at least for an example of two pendulums: no matter how you start their motion, the resulting pattern is always a proper combination of

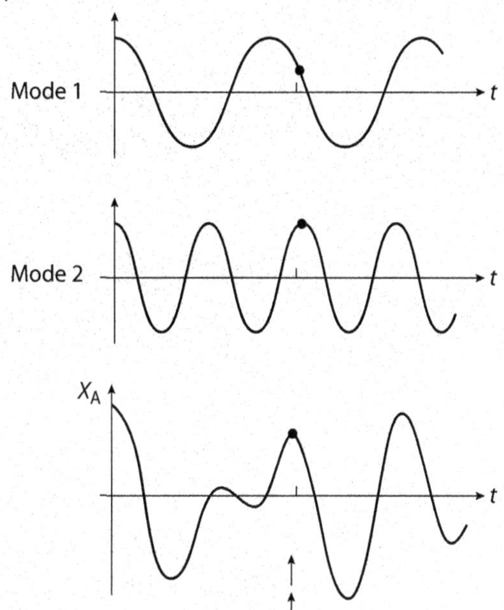

Fig. 10.5. First mode (upper graph) and second mode (in the middle) being added are giving the non-periodic function (lower graph).

Adapted from Musical Acoustics by Donald Hall, 2002 Cengage Learning, Inc.

two normal modes resulting in seemingly complex motion. There is a strict mathematical procedure that allows us to find the content of normal modes in any complex motion, but we will continue the consideration conceptually.

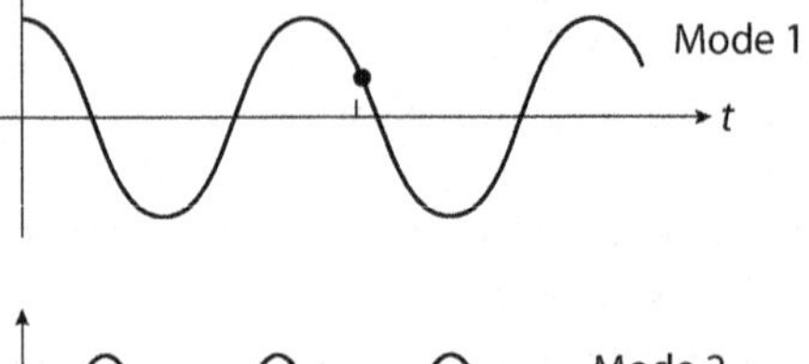

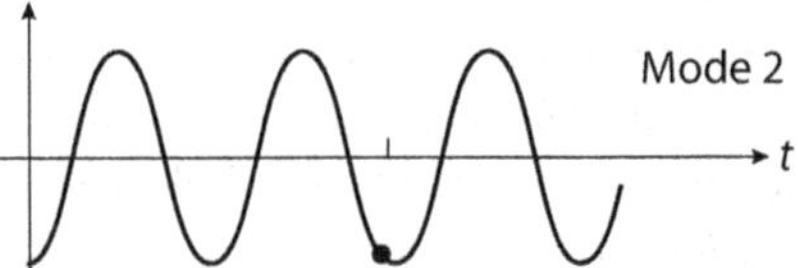

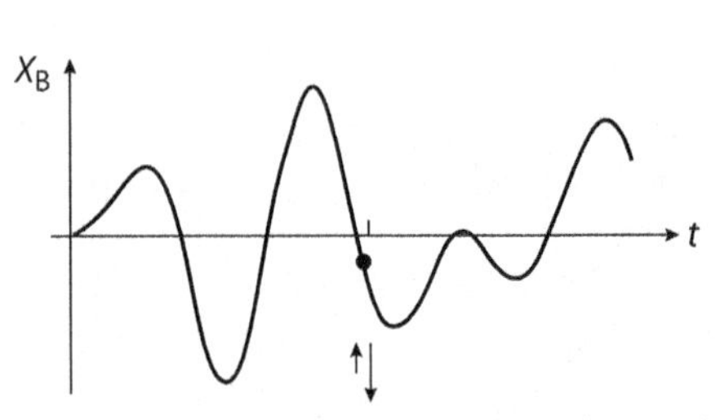

Fig. 10.6. First mode (upper graph) and second mode (middle graph) being added are given the non-periodic function (lower graph).
Adapted from Musical Acoustics by Donald Hall, 2002 Cengage Learning, Inc.

10.3 Normal Modes for Systems with More Degrees of Freedom

Two pendulums are pretty far from a tuning fork, whose motion we want to explain. So, we make another step.

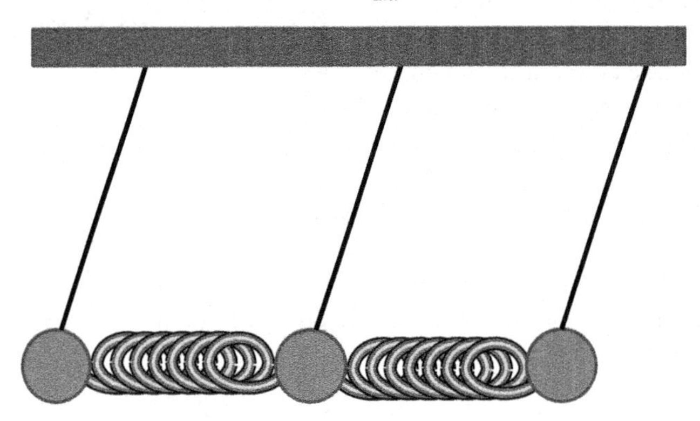

Fig. 10.7a. First mode of the system of three pendulums connected by springs.

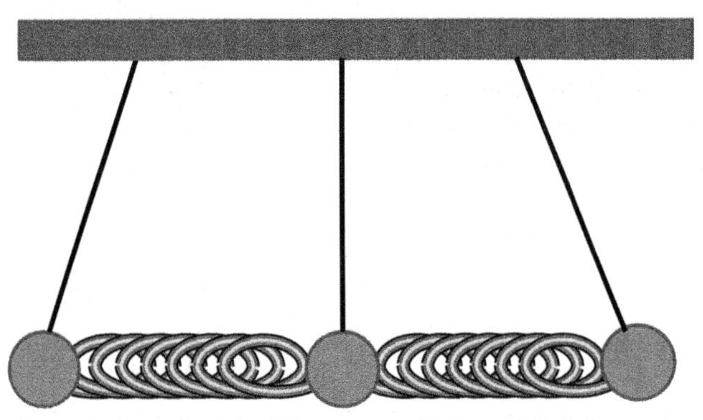

Fig. 10.7b. Second mode of the system of three pendulums connected by springs.

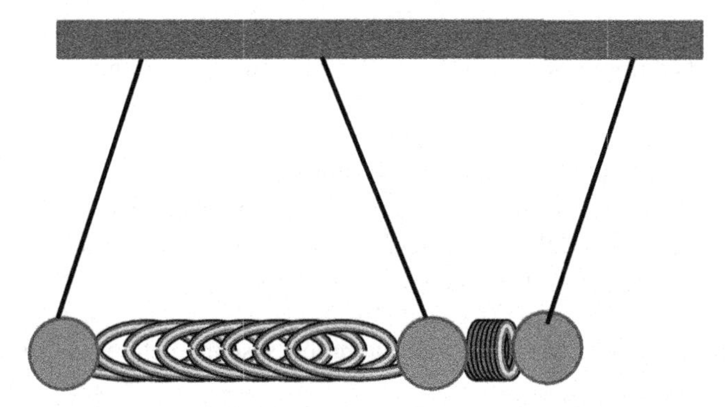

Fig. 10.7c. Third mode of the system of three pendulums connected by springs.

Let us consider systems of three, four, five pendulums and try to extract some general important features. For three pendulums it is pretty easy to prove that we have three types of initial conditions that will lead to the SHM motion of a system (Fig. 10.7). Again,as in a situation with two pendulums, the mode with the lowest frequency will appear when we start motion by pulling all three pendulums at the same distance to the same side. Again, the springs do not participate in motion, and three pendulums demonstrate the same frequency as one pendulum alone (Fig. 10.7a). The second mode corresponds to the initial conditions, when the first and third pendulums are pulled at the same distance in opposite directions; meanwhile, the pendulum in the middle stays at rest. The energy of motion now is stored in two pendulums and two springs (Fig. 10.7b). The third mode starts with the initial conditions when one spring is compressed and another stretched with the same elongation. Now all three pendulums participate in the

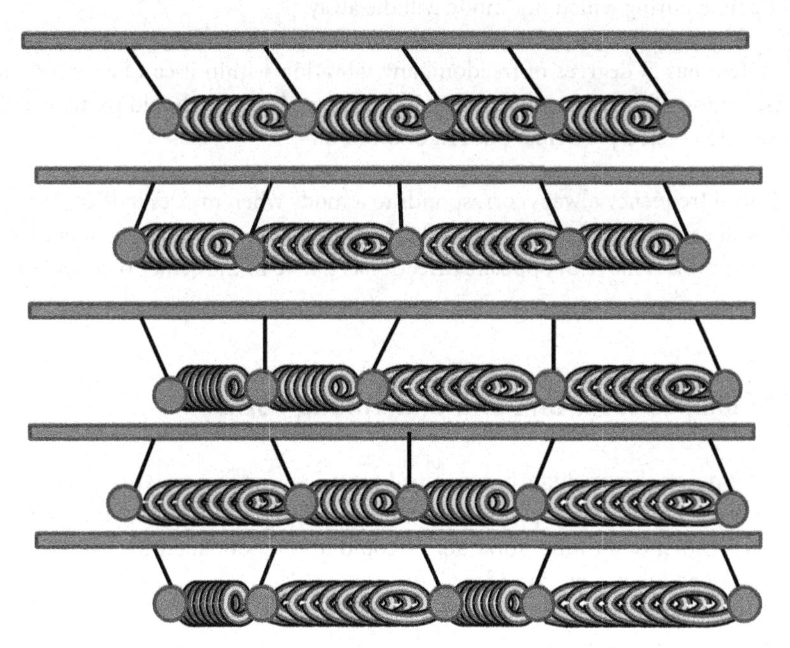

Fig. 10.8. Five normal (natural) modes of the system of five pendulums connected by springs.

motion (Fig. 10.7c). For a system of three pendulums, we have three different normal modes with frequencies f_1, f_2, f_3 from lowest to highest. These frequencies may not be the members of a harmonic series.

Normal modes for a system of five pendulums are shown in Fig. 10.8. It is possible to predict that this system will demonstrate **five** normal modes with different frequencies. The lowest mode corresponds to motion of pendulums "in step," the next one, when two left and two right are moving in opposite directions while the center one is at rest.

Let us summarize the properties of abstract vibration systems before we can approach the realistic tuning fork:

- We use the letter N to denote the **degrees of freedom** that exist in every system of vibrating masses This number determines the number of possible independent SHMs. There are never more than N ways to set a system into SHM.

- If a system has N degrees of freedom, this means it has exactly N normal modes: never fewer, never more. These modes establish a motion pattern so that every part of our system moves in SHM. Within a particular mode, a point that continues at rest (such as the center pendulum for a second mode in a system of three or five pendulums) is called a **node**.

- Usually, all N frequencies are different, and normal modes have their own characteristic frequencies. Each frequency is determined by the amount of potential energy needed to be stored in our system to excite this particular mode: the sum of potential energies of gravity and spring constants of participating springs. This energy also defines the characteristic time during which our mode will die away.

- If a system has N degrees of freedom, any vibration within it can be represented as the superposition of the N normal modes. These normal modes should participate in proper amounts to create a particular pattern of motion.

- The lowest frequency always corresponds to a mode when masses in the system move "in step" with each other. The second mode corresponds to the situation when left and right groups of masses move in opposite directions relative to the center of a system.

10.4 Bars Clamped from One End and Tuning Forks

Let us consider now a thin metallic bar clamped from one end, for instance, secured somehow to the wall (Fig. 10.9a). This example is important from two perspectives: it represents a simplified model of a prong of a tuning fork, and it could also be obtained by a simple increase in the number of pendulums as considered in the previous subchapter. How do we go from a set of pendulums to a metallic bar? For this we should just recall that matter consists of tiny atoms (masses) kept together by springy interatomic forces (effective springs). The only

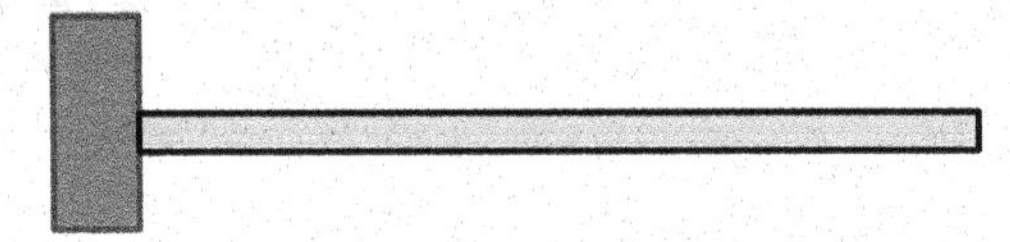

Fig. 10.9a. Bar clamped at one end, at rest.

problem is that now we have, say, 10^{23} masses, which means 10^{23} normal modes. It is impossible to work with such a number of modes, moreover, it does not make any sense.

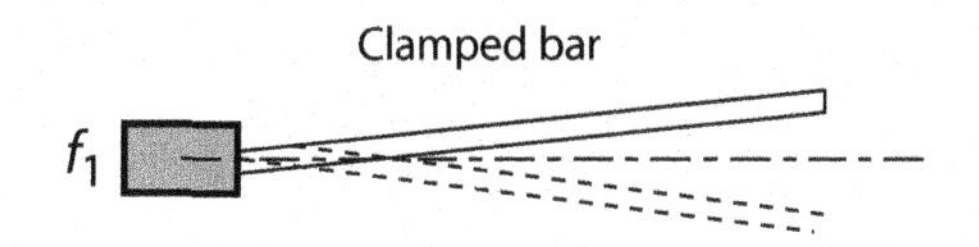

Fig. 10.9b. First mode of the bar clamped at one end.
Adapted from Musical Acoustics by Donald Hall, 2002 Cengage Learning, Inc.

The conclusions of the previous subchapter help us to predict the character of the first fundamental mode, exactly like the example with several pendulums, the very first mode, which corresponds to the lowest frequency, which will occur when all the little masses sitting inside of our bar are moving "in step" (Fig. 10. 9b). Because the bar is clamped at one end, the amplitudes of these masses are not the same, gradually decreasing to zero at the secured end, but all the masses are moving "in step." The frequency of this mode depends on a value called Young modulus (effective springiness of a bar) and density, which means mass per unit volume of a bar. Both of these quantities are completely defined by the material of which this bar is made. The big influence on the normal frequencies is also the length of a bar, which will be discussed later.

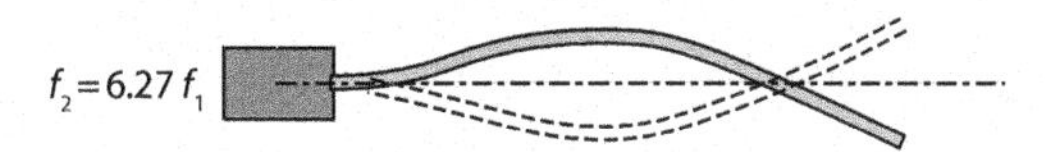

Fig. 10.9c. Second mode of the bar clamped at one end.
Adapted from Musical Acoustics by Donald Hall, 2002 Cengage Learning, Inc.

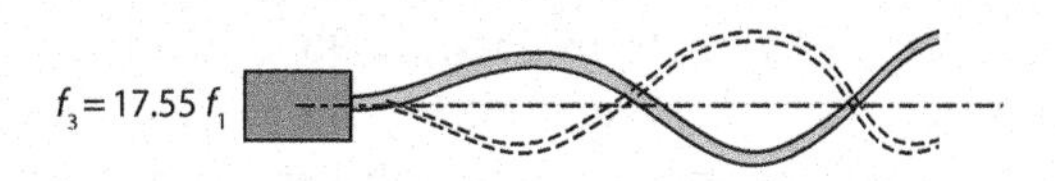

Fig. 10.9d. Third mode of the bar clamped at one end.
Adapted from Musical Acoustics by Donald Hall, 2002 Cengage Learning, Inc.

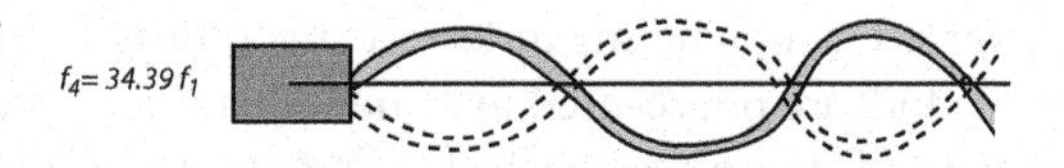

Fig. 10.9e.
Adapted from Musical Acoustics by Donald Hall, 2002 Cengage Learning, Inc.

The second mode should somehow have parts of a bar moving in opposite directions. Because our bar is clamped, the nodal point is not exactly in the middle but is closer to the "free" end (Fig. 10.9c). Next, two normal modes of a bar are shown in Figs. 10.9d and e. The frequencies of each mode are indicated.

The general formula for a fundamental frequency of a clamped bar is

$$f_1 = \frac{0.162a}{L^2}\sqrt{\frac{Y}{D}}$$

where a is the thickness of a bar; L, the length of a bar; Y, the Young modulus; and D, the density of material of which the bar is made. The frequencies of upper modes can be obtained from the first one by a recurring formula:

$$f_n = 2.81\left(n - \frac{1}{2}\right)^2 f_1$$

Note that the fundamental frequency of a bar depends on the length as $1/L^2$. This means that if you cut the bar twice, the fundamental frequency will increase four times.

Example 10.1. You want to create your own musical instrument based on clamped bars. What should be the ratio of length of otherwise identical bars to create a difference in pitch of 1 octave?

Answer:

The ratio of frequencies is equal to 2. Frequency depends on the length indirectly proportional to L^2. So,

$$f_2/f_1 = 2 = \left(\frac{L_1}{L_2}\right)^2, \text{ or } \frac{L_1}{L_2} = \sqrt{2} = 1.41.$$

For two bars to have a difference in pitch of one octave, one bar should be longer than the other by 1.41 times.

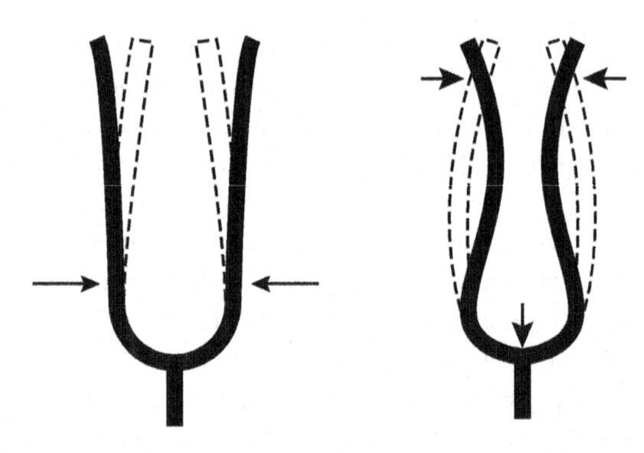

Fig. 10.10. Two first modes of the tuning fork.
Adapted from Musical Acoustics by Donald Hall, 2002 Cengage Learning, Inc.

Another property of a simple metallic bar clamped from one end is a strong inharmonicity of its spectrum. The second, third, and fourth modes relate to the fundamental, as can be seen from Figs. 10.9b–e, with ratios 5.27, 17.55, and 34.56. These frequencies do not create harmonic series. This bar when struck will create a disturbing, dissonant sound.

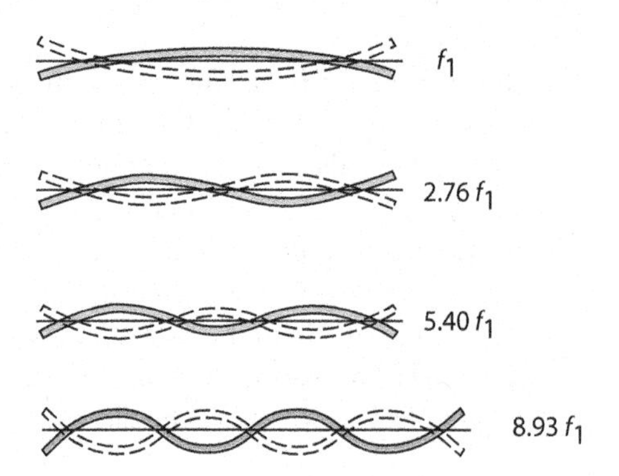

Fig. 10.11. First four modes of the free bar.
Adapted from Musical Acoustics by Donald Hall, 2002 Cengage Learning, Inc.

Now let us apply all we have discussed in this section to a prong of a tuning fork. Sure, the prongs of a tuning fork are not secured to some wall; rather, they are embedded into a base, which is slightly movable itself. As a result, the second mode of a tuning fork is close to the ratio 5.1 with a fundamental. The motion of the prongs of a tuning fork in the first two modes is shown in Fig. 10.10. By proper design of a base, the first two modes can give the ratio of frequencies $f_2/f_1=6$ so that the clang tone for which the second mode is responsible creates a decent musical interval with fundamental in two octaves and a perfect fifth (2x2x1.5). The third and higher modes of a tuning fork are of little musical interest. First of all, they have very high frequencies so that fifth mode will be beyond the audible range, and second, they die away very quickly. The third mode is responsible for a metallic ping when the fork is struck, but only the first and second modes give more sustained tones.

10.5 Free Metallic Bars and Xylophone Bars

Now let us consider wooden or metallic bars, such as in xylophones. The difference between a xylophone and a prong of a tuning fork is that xylophone bars are free from both ends.

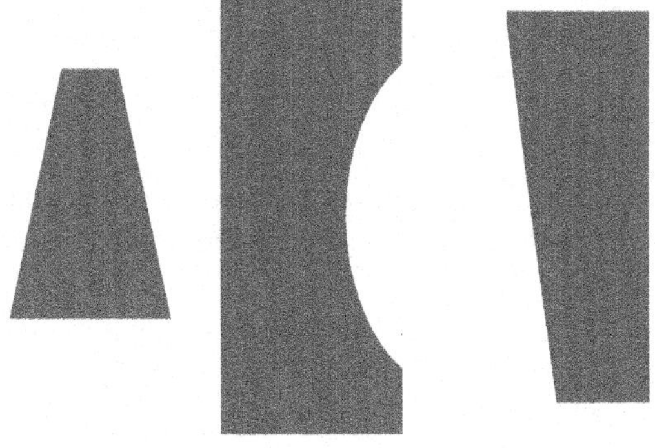

Fig. 10.12. Forms of metallic bars which can have harmonic series in their spectrum.

The first four modes of a bar free from both ends are shown in Fig. 10.11. The fundamental frequency of such a bar depends again on the thickness a, length of a bar L, and properties of a material Y and D:

$$f_1 = \frac{1.028a}{L^2}\sqrt{\frac{Y}{D}}$$

The higher mode frequencies are given now as

$$f_n = 0.441\left(n + \frac{1}{2}\right)^2 f_1$$

The second, third, and fourth modes have ratios with fundamental 2.76, 6.4, and 8.93, which again is far from any harmonic series. The ratio 2.75, for example, is 17.5 semitones apart, which is not very pleasant musically at all. As a result, the bars used in xylophones are usually not perfectly uniform, for example, like several shown in Fig. 10.12. For such bars the second mode creates with a fundamental the ratio 4, which are perfect 2 octaves, and these two modes reinforce each other. Note that the thinning of a bar in the center drastically reduces the stiffness of the bar, lowering the frequency of its first normal mode. The higher modes are left almost unaffected because their creation still requires the bending of thick peripheral parts. Such thinning demonstrates bars used in xylophones and marimbas. Marimbas' bars have ratio $f_2/f_1=4$, xylophones usually use $f_2/f_1=3$.

Example 10.2. Two bars are made of the same material. One has the length twice the size of the second and a cross-sectional area three times as big as the second. What is the ratio of frequencies of their fundamental modes? Is there any difference in this ratio if these bars are clamped or free at both ends?

Answer:

The frequency of a bar depends on its length and cross-sectional area as a/L^2. We have two bars: one of frequency f_1 with length L and cross-sectional area a, and another of frequency f_2 with length $2L$ and cross-sectional area $3a$. The ratio of frequencies can be written as

$$f_1/f_2 = \left(\frac{a}{L^2}\right) \cdot \left(\frac{4L^2}{3a}\right) = \frac{4}{3}$$

10.6 Vibration of Surface: Drums and Cymbals

How can we represent a drumhead attached to an isolated hoop? Following the procedure developed above, we can describe such a system as an array of masses connected to each other with effective springs (Fig. 10.13). Again, the number of modes will be equal to the number of degrees of freedom, which means the number of masses. The first mode again corresponds to motion of all masses "in step" (Fig. 10.14a). With upper modes everything becomes a little bit more complicated than for a linear bar. All we know for sure is that all upper modes will have nodes, but in two-dimensional situations it is better to say that modes will have **nodal** lines. The parts of a membrane on different sides of a nodal line are moving in opposite directions, meanwhile, the

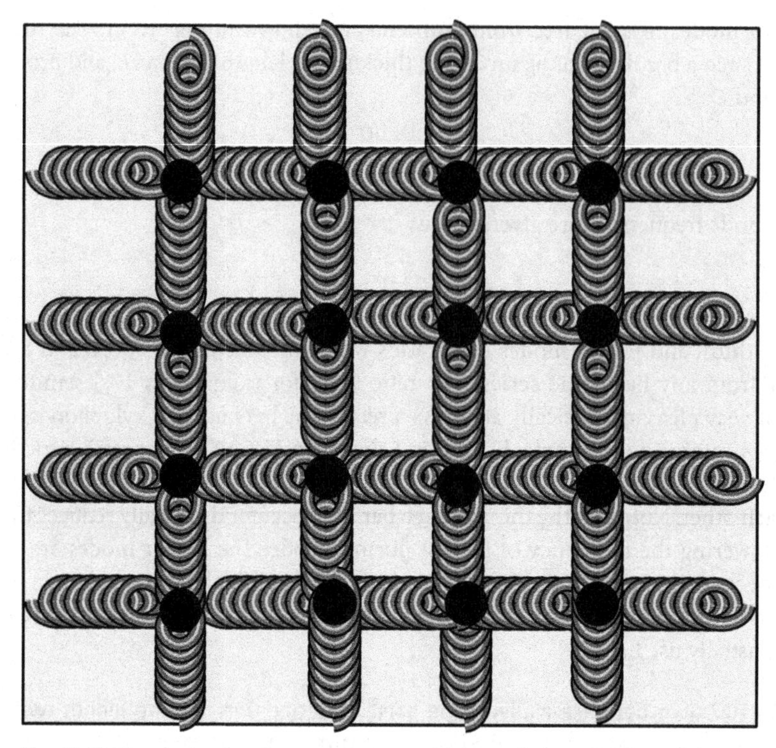

Fig. 10.13. The array of spring and masses which can be used as a model of a drumhead.

nodal line remains at rest. To make Fig. 10.14 clearer, we are using the sign "+" to indicate the area that is moving toward us at a given moment and the sign "–" for a part that is moving away.

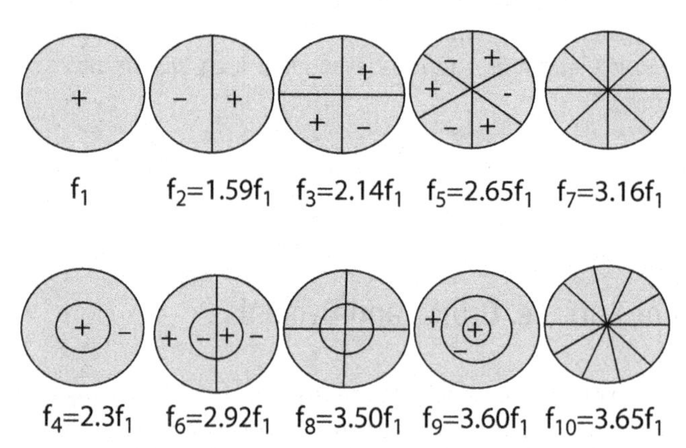

Fig. 10.14. Modes of the circular drumhead. Frequencies of modes are shown.
Source: Musical Acoustics by Donald Hall, 2002 Cengage Learning, Inc.

Now we can analyze all 10 modes shown in Fig. 10.14. First of all, there are two different types of modes. The first type has a nodal point at the center and looks like some sort of pie chart. Such kinds of modes are called radial modes (see, for example, modes 2, 3, 4, 6, 7, 8, 10). Another type

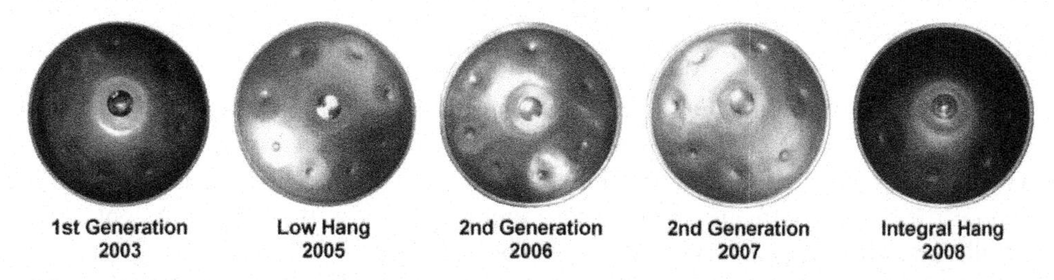

Fig. 10.15. Hang drums.

of mode has nodal lines in the form of circles, so the nodal line never goes through the center of our membrane (modes 1, 4, 9). These modes are called circular modes. To excite these two different types of modes we should strike our drumhead at different points. Mode 1 pushes all the air contacting with the membrane in the same direction, radiating the sound energy very efficiently; as a result, it loses energy really fast and disappears from the later persistent sound.

The drumhead vibration does not generate musical pitch because the frequencies' ratios do not fall into musical notation. The resulting sound of a drumhead is a mixture for which it is practically impossible to assign any musical pitch. Addition of a body to a membrane can create resonating frequencies of a body, which alters the efficiency with which modes radiate. The special form of a body can make certain mode frequencies harmonic. The percussion instrument called a hang drum demonstrates a nearly harmonic spectrum and is shown in Fig. 10.16.

10.7 Striking Points and Following Vibration

Now we should answer a very important question: how can we set a drumhead into vibration in a single pure normal mode? There is only one answer:

We force the different parts of a membrane to take a form corresponding to the mode needed. For instance, to set a vibration in mode 2, we should pull half of the membrane up and half of it down. But nobody does it, first, because it is extremely difficult (even for a mode with one nodal line), and second, because nobody uses a drumhead to work like a tuning fork: the drumhead has its own voice. Usually we strike a membrane, at a small point in comparison to the whole area, with a hammer or a mallet.

Striking a drumhead at its exact center yields only a muffled thud. This is because at the center, you can excite only circular modes, leaving outside the vast majority of radial modes. To make the radial modes strong you should hit the drumhead approximately one-half to three-quarters of the distance from the center to the edge.

The same set of modes is produced when you strike an entire region at once (as opposed to a single point) as when you add together the modes at each point of this region. Note that if the region of striking includes both "+" and "–" areas for a mode, such influence can kill or, at the very least, drastically decrease the participation of that particular mode.

Another important factor that influences the voice of the drumhead is the time of contact of a mallet with the membrane. The longer this time is, the lower the pitch the drumhead demonstrates. This can be understood from Fig. 10.15. It the very first moment the membrane is set into motion downward (Fig. 10.16a) in all possible modes, which do not have a nodal point at the point of contact. But after half of the period of the corresponding mode, this part of a membrane moves up (Fig. 10.16b). If the mallet is not removed, it will kill the just-excited vibration. The higher the frequency of the corresponding mode is, the shorter the contact time of a mallet and a drumhead must be. All this can be summarized into:

Only modes whose frequencies are less than *2/T* are efficiently excited when a striking force has a finite duration of *T*.

Using a soft mallet or hard hammer definitely produces a difference in the character of the sound. The hard hammer applies great pressure to a small area, exciting many modes at this point and producing bright and full sound. The soft mallet exerts lesser pressure over the bigger area, so the modes that have nodal lines under the mallet are removed from the equation. As a result, it produces lower frequencies. Also, since the soft mallet bounces back more slowly, the contact time is greater, which lowers possible frequencies even more.

If we strike a membrane in the familiar way, it excites the modes whose patterns are the closest to the force patterns. This means that all modes that have moving points at the striking point will be excited. Meanwhile, modes that have a nodal line at the given striking point will be completely indifferent to our efforts: this line does not move. This mode "does not know" that it's time to play.

The closeness of the mode pattern to the pattern of a striking force determines how each mode is excited. Striking an object at any given point excites each natural mode in proportion to how much that mode moves at this particular point.

For example, we cannot excite any radial (pie-like) modes by striking a drumhead exactly in the center. All these modes do have a node in the center, and they simply will not "know" that we want them to move.

So, **if the striking point falls on a nodal line or point of a drumhead, the corresponding mode is not excited and is not present in the produced sound.**

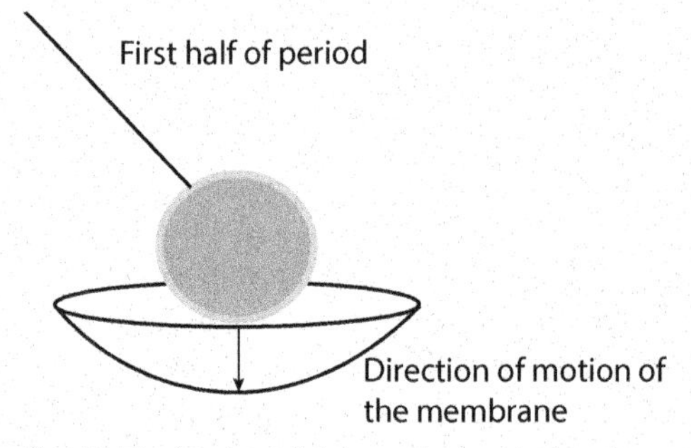

Fig. 10.16a. Motion of the membrane during the first half of period of oscillations.

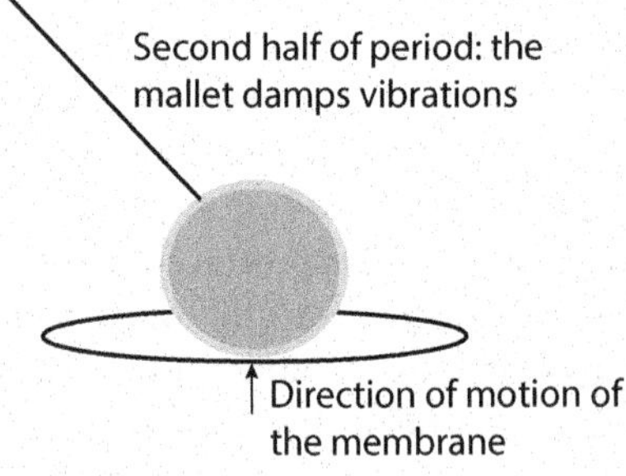

Fig. 10.16b. Motion of the membrane during second half of period of oscillations: the mallet is killing the vibration.

10.8 Damping Times

When the drumhead vibrates, it transfers energy, setting into motion the adjacent areas of air. As a result, the energy flows away from the vibration, and the sound dies. So, percussion instruments always demonstrate the damped vibrations whose amplitudes decrease with time.

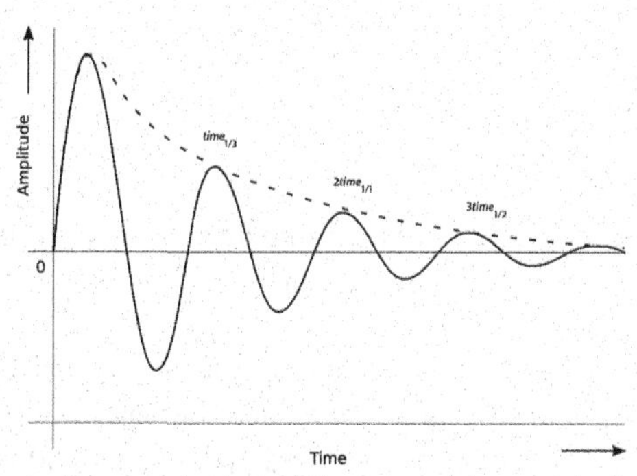

Fig. 10.17. Dependence of the amplitude of vibration on time. The characteristic halving time is shown.

The simplest form of damping is exponential damping when the rate of loss of energy is always proportional to the energy remaining in a system. The characteristic of such a dumping, called the halving time ($t_{1/2}$), is a time during which the initial amplitude of given vibration decreases twice relative to the initial value. This process is illustrated in Fig. 10.17: after the first halving time the amplitude of original vibration is decreased twice, after second halving time, again twice, or four times relative to the original value; after three halving times it will decrease twice again, or eight times in comparison to the original value.

For musical purposes it is useful to define the damping time τ (tau) of any mode as the time during which the amplitude of the mode is falling to 1/1000 of original value. From previous chapters we know that a 1/1000 decrease in amplitude corresponds to the 1/1000000 decrease in intensity and 60 dB drop in SIL. A 60 dB drop usually fades a signal below any noise levels. The correspondence between damping time and halving time is $\tau = 10\ t_{1/2}$.

Damping times are different for different modes. Any mode with a more efficient sound radiation has a shorter damping time. The high-frequency modes of metallic bars usually have very short damping times because they involve sophisticated and highly energetic bending. We can manage the difference between damping times by the application of localized frictional damping. For instance, touching the drumhead exactly at the center will quickly kill all the circular modes, leaving other modes that have a nodal point at the center undisturbed.

The variation of damping time for different modes can cause the change of the tone color during decay. Just after striking there are a lot of the highest modes, which are dying away one after another. The combined frequency of the sound is going lower and lower, becoming more simple and dull.

Summary, Terms, and Relations

The pattern of motion that gives SHM is called normal mode (also natural mode), and each pattern has its own normal mode (natural mode) frequency.

Each system with N degrees of freedom must have exactly N normal modes: never fewer or more. Each normal mode has its own characteristic frequency, and usually all N frequencies are different.

The fundamental frequency of the metallic bar depends on the length as $1/L^2$.

Striking an object at any given point excites each natural mode in proportion to how much that mode moves at this particular point. If the striking point falls on a nodal line or point of a drumhead, the corresponding mode is not excited and does not participate in the recipe of the produced sound.

If a striking force has finite duration *T* in time, then only modes whose frequencies are less than $2/T$ are efficiently excited.

The halving time ($t_{1/2}$) is a time during which the initial amplitude of given vibration decreases twice relatively initial value. The damping time τ (tau) of any mode is the time during which the amplitude of the mode falls to 1/1000 of its original value. The correspondence between damping time and halving time is $\tau = 10\ t_{1/2}$.

Questions and Exercises

1. Sketch normal modes for a system of four pendulums connected with springs. Assign mode numbers from lowest to highest frequency.

2. What modes will be excited as you strike the xylophone bar exactly at its center?

3. Discuss how a xylophonist can obtain a different timbre by striking the bar either in the center or at different distances from the center.

4. Suppose you have a bar clamped from one end, whose fundamental mode corresponds to the note A_4. What fundamental will you get if you cut this particular bar into halves?

6. What modes will be excited on the circular drumhead if you strike it exactly at the center?

5. If the amplitude of some mode dies to half of its original value in 0.1 s, what is its 60 dB damping time?

7. The fundamental frequency of a bar that is free at both ends is 100 Hz. What will the frequency be of another bar that is made of the same material and of the same thickness but is one third as long?

8. Two bars, A and B, are made of the same material and have the same length. But bar A is twice as thick as bar B. Which of these bars will have a higher fundamental frequency? What is the ratio of these frequencies? Does your answer depend on whether these bars are clamped at one end or free at both ends?

9. Suppose a drumstick remains in contact with a wood block for 0.4 s. At roughly what frequency is the dividing line between efficient and inefficient mode excitation?

10. Discuss how you can damp out the first mode of the drumhead.

CHAPTER 11

Normal Modes of Strings

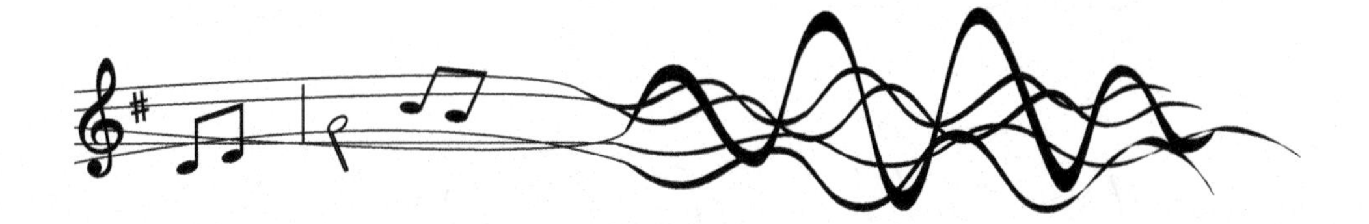

11.1 Standing Waves on Thin Strings

Nature gives us several objects, which by themselves demonstrate the harmonic spectra. This means that the normal modes of such vibrating objects create the harmonic series. Thin strings used in guitars, violins, harps, and pianos are one of them. No wonder that humans used strings for very important musical purposes: to build a family of instruments.

What is a string from the point of view of Physics? First of all, it is a flexible and stretchable object, which can be considered a very thin metallic bar. So thin that we can neglect the cross-sectional area and consider a string as a one-dimensional, linear system. Second, string is always stretched tightly between two supports, as shown in Fig. 11.1. As a result, the end of the string cannot move, naturally creating nodes. One of the supports is called a **bridge**, and it is mounted on a soundboard; another one is either a **nut** and is mounted on the heavy frame of a piano or at the end of a guitar or violin neck, or one of the **frets** on a guitar fingerboard.

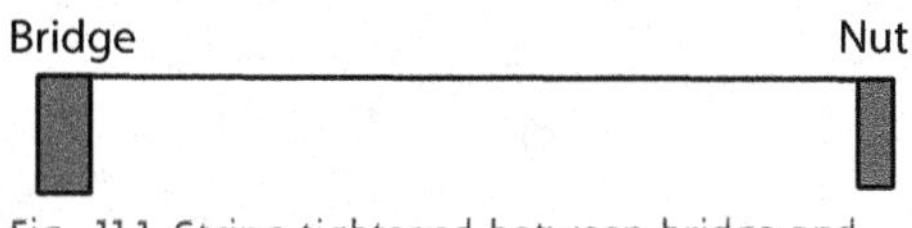

Fig. 11.1. String tightened between bridge and nut, at rest.

To visualize the creation of normal modes on a thin string, let us consider two girls playing with a jump rope (Fig. 11.2). When the girls rotate a rope with small frequency, it behaves as shown in part (a) of the figure. This form looks like a part of a wave. But the ends of the rope are secured, so this wave cannot travel through the ends. Moreover, this wave does not travel at all: the point exactly at the middle always oscillates back and forth with the biggest amplitude; meanwhile, points at the end are always at rest. Such a wave, which does not travel, is called a **standing**

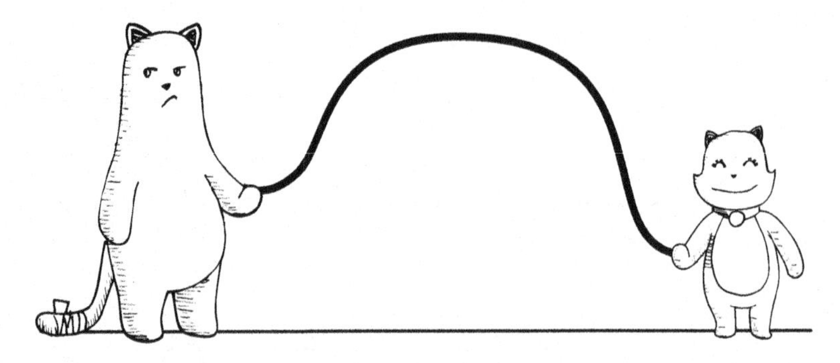

Fig. 11.2a. The first mode on the string. Here a jumping rope is shown as an example.

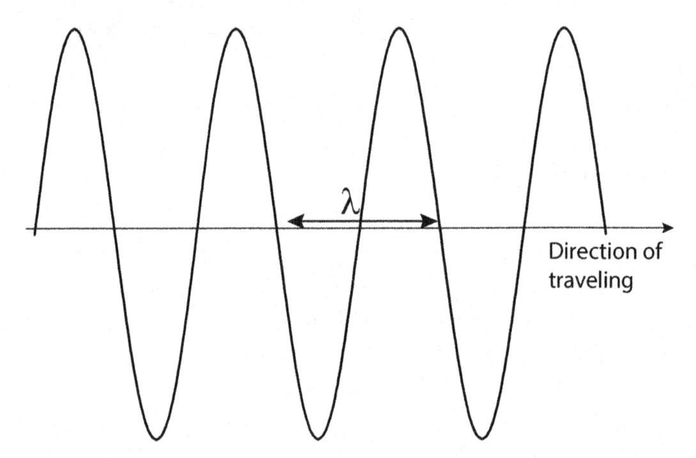

Fig. 11.2b. Traveling wave: the wavelength is shown.

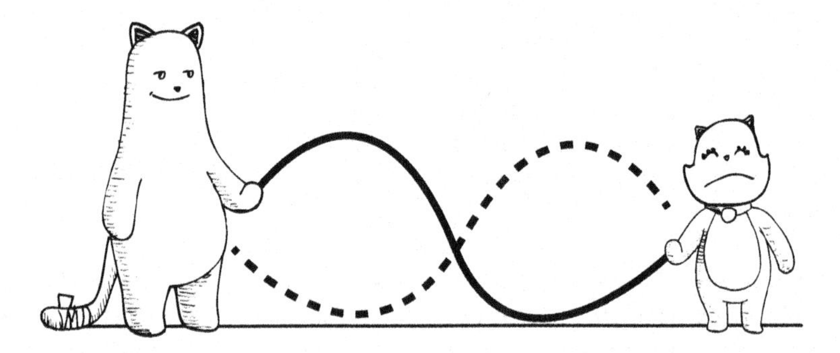

Fig. 11.2c. Second mode on the rope (string) appears when the frequency of excitation is increased.

wave. The point with the biggest amplitude of vibration is the **antinode**, and in the example we consider here, it is exactly at the center of the rope.

How is a standing wave created and where does it come from? The standing wave is a result of the superposition of two traveling waves, one that falls on the support, and the other is reflected from it. The process of adding two waves, one traveling to the right and the other to the left, is shown in Fig. 11.3. Because a standing wave is a result of simple superposition, all the properties of traveling waves, such as mutual correspondence of frequency, wavelength, and speed of wave are valid for the consideration of standing waves too.

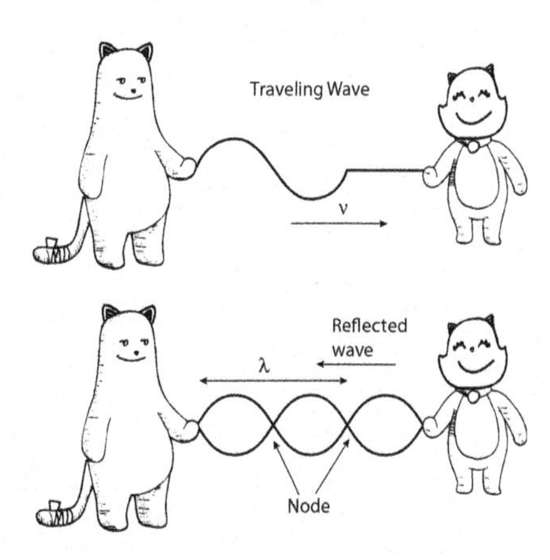

Fig. 11.3. Standing wave as a result of superposition of traveling wave (upper part) and reflected wave.

Because we have been prepared by a previous chapter, we already expect that the standing wave, shown in Fig. 11.2a, is not the only normal mode of a string. How we can predict all others? For this check Fig. 11.2b. The first mode is created by a piece of traveling wave between two adjacent zeros. To be exact, this is precisely half the wavelength of a traveling wave, from which our standing wave has been created. The next normal mode will occur when the length of rope fits the distance between one zero on a traveling wave and non-adjacent subsequent one. Or when the length of rope is exactly equal to the wavelength of a traveling wave. The girls in Fig. 11.2c increased the frequency of rotation of a rope, and, as a result, we see the second mode. This standing wave does not move anywhere, by definition, and the difference between the first and second mode is that now we have two antinodes (at a distance of one quarter of the total length from each end) and the node in the center.

Now let summarize all conclusions. The first normal mode on the string fits the half of a wavelength of a traveling wave:

$$L = \frac{\lambda_1}{2}$$

The frequency of vibration is connected with a wavelength through a speed of the wave:

$$f_1 = \frac{V}{\lambda_1} = \frac{V}{2L}$$

The second mode appears when the wavelength of a traveling wave fits exactly the length of the rope:

$$L = \lambda_2$$

$$f_2 = \frac{V}{\lambda_2} = \frac{V}{L} = 2f_1$$

These two frequencies have the ratio 2 and belong to the same harmonics series! The next mode will fit one and a half of a wavelength of the traveling wave, etc. The first four normal modes of a string are shown in Fig. 11.4. Now we can write the general formula for a frequency of a normal mode number n on a thin string:

$$\lambda_n = \frac{\lambda_1}{n}$$

$$f_n = \frac{V}{\lambda_n} = n\frac{V}{2L} = nf_1$$

where n is any integer. We may also make a conclusion that the normal mode number *n* has *n–1* antinodes and *n+1* nodes (the edge nodes included).

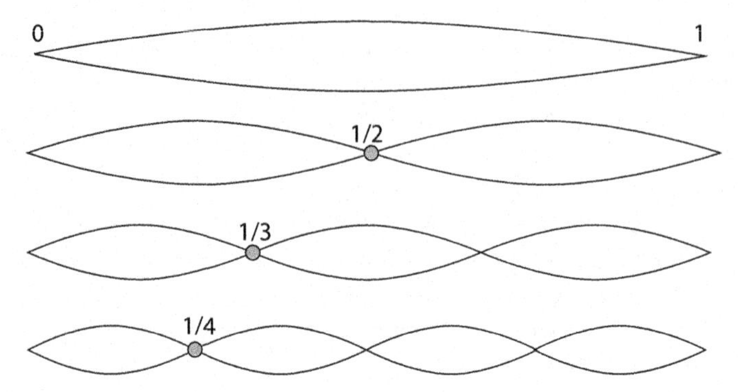

Fig. 11.4. First four normal (natural) modes of the thing string. Positions of nodes are shown as fraction of length of the string.

Example 11.1. The string is 1 m long. What are the wavelengths of the first 3 normal modes on this string?

Answer:

The wavelength of the nth mode can be found using

$$\lambda_n = \frac{2L}{n}$$

For n=1, λ_1= 2 x 1m = 2m.; for n=2, λ_2= 2 x 1m/2=1m; for n=2, λ_3= 2 x 1m/3= 0.66 m.

Example 11.2. The string is 1 m long. What is the distance of the first antinode from a bridge for the three lowest modes?

Answer:

The first mode has an antinode exactly at the center of a string, L/2 = 0.5 m from a bridge. The second mode has the antinode (refer to Fig. 11.2c) at the distance L/4 = 0.25 m from a bridge; the third mode has an antinode at the distance L/6 = 0.166 m from a bridge.

Example 11.3. What is the number of nodes and antinodes for seventh mode of the string?

Answer:

The seventh mode has 8 nodes and 7 antinodes.

It is important to understand that the speed of wave, which we used in previous formulas, is not the speed of sound in the air. This is the speed of travel in the string wave. It depends on the force of tension of a string and mass per unit length, as we already know from earlier chapters of this book:

$$v = \sqrt{\frac{F_t}{\left(\frac{m}{L}\right)}}$$

Typical values of force of tension for a guitar (D string) are 150 N and mass per unit length 0.005 kg/m, which gives a speed of around 170 m/s; for a piano, force of tension is around 600 N for middle range and mass per unit length is 0.006 kg/m, with typical speeds around 330 m/s.

Example 11.4. The speed of a wave in a string is 180 m/s and the length of this string is 120 cm. What are the frequencies of the first 5 modes in this string?

Answer:

The frequency of fundamental mode is $f_1 = v/\lambda_1 = v/2L = 180\ m/s/(2x1.2\ m) = 117\ Hz$. The frequencies of upper harmonics can be obtained by simple multiplication of f_1 by integer: $f_2 = 2 \times f_1 = 234\ Hz$, $f_3 = 3xf_1 = 351\ Hz$, $f_4 = 2xf_1 = 468\ Hz$, $f_5 = 5 \times f_1 = 585\ Hz$.

Example 11.5. How will the pitch change if we double the length of string, leaving all other parameters unchanged?

Answer:

The frequency of the fundamental mode depends on the length of the string as 1/L. This means if we increase the length twice, the frequency will decrease two times, and the pitch will drop by an octave.

11.2 Plucked Strings

11.2.1 Plucking Points on Guitars

When we are talking about girls with a rope, we can discuss the possibility of the creation of a single standing wave. But when we pluck a string of a real guitar, the picture becomes more sophisticated. First of all, we do not create a wave from one end; we pluck the string at some arbitrary point, usually at the sound hole of the guitar's resonator. The result of such an action is a couple of pulse waves, usually called **kinks** (Fig. 11.5a). Created kinks move in opposite directions, reflect from the corresponding supports, and continue this bouncing back and forth.

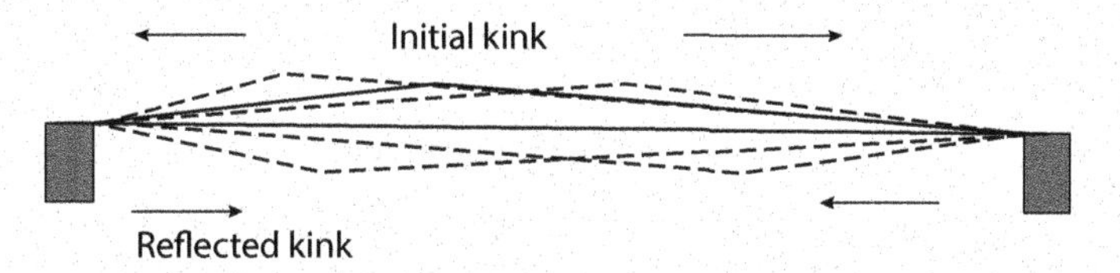

Fig. 11.5a. Traveling of initial kink on the string.

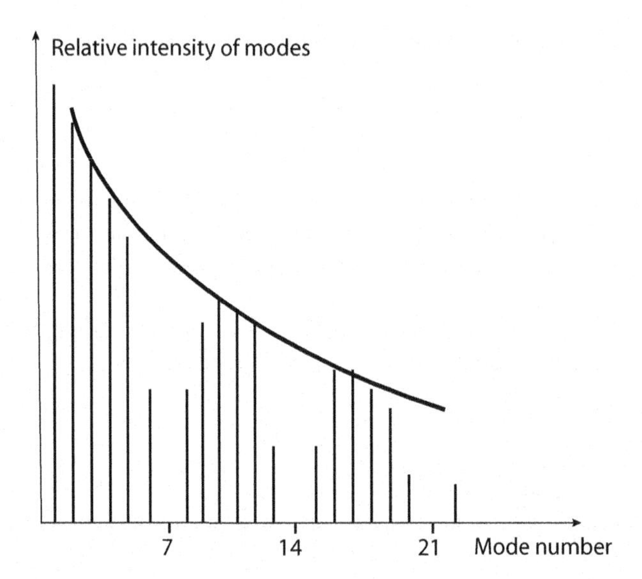

Fig. 11.5b. Spectrum of sound created by plucked string. The plucking point is L/7.

The form of kink is something between a perfect triangle and sawtooth waves. And we know that such a waveform creates all sets of upper harmonics. The amplitude of these harmonics falls at the rate of 6 dB per octave.

But the question is: are we exciting all the upper harmonics? Does the point of plucking influence the tone color? Actually, we already know the answer from a previous chapter. The excitation of any mode is proportional to how much motion this mode has at a given point, meaning that mode with a node at the plucking point will be completely left out of the created sound. Fig. 11.5b shows a spectrum of a sound created when a string was plucked at distance L/7 from one of the ends. As a result, modes 7, 14, 21 and all others whose numbers are multiples of 7 disappear from tone color because they all have nodes at the given point. The worst situation occurs when we pluck a string exactly at the middle. All the even-numbered modes have the node at the center of the string, and they all will be omitted from our music.

Example 11.6. The plucking point on some guitars is 1/8 of the length of a string. What modes will disappear from the spectrum of a string?

Answer:

All modes that have a node at the point of plucking will disappear from a sound. At distance L/8 from either end, there is nodal point for modes 8, 16, 24, 32. They will be not excited.

11.2.2 Influence of Pickup Position

The sound of a string alone does not correspond to what you hear because the vibration should be transmitted to the air first. The acoustic guitar has a box-like resonator with a circular hole to increase the mutual connection of the vibrations of the air inside the body with the air outside, which delivers sound to our ears. The body of an acoustic guitar, especially the upper plate, must be pretty light to maximize its response to the string motion. The hollow cavity inside the resonator assists to the lower pitches.

The body of an electric guitar is just a rigid platform. This results in different times of decay of sound for acoustic and electric models. Another important problem, which an acoustic guitar simply does not have, is the position of the pickup point of an electric guitar.

How does a pickup point work? In a nutshell, a pickup point is a small magnet mounted underneath the strings, and the metallic string (for electric guitars we use only metallic strings) is the moving magnetic field of this magnet. The effect, which is used in electric guitars, is the same as the one used at checkpoints in airports when you move through a frame. If you have a piece of metal in your pocket, it disturbs the magnetic field inside this frame, and you hear a signal. Same with the electric guitar: moving inside the magnetic field string disturbs the magnetic field, and this signal can be amplified and processed. But for amplification of a signal, the string must move. If some mode has a node at the pickup point, this mode simply can't be amplified: a magnet does not know anything about the existence of this mode. So, **each normal mode is amplified by the pickup in proportion to how much motion that mode produces at the pickup location.**

If we have a pickup point at L/13, the modes 13, 26, and others whose numbers are multiples of 13 will disappear from the amplification. Fig. 11.6 illustrates the spectrum of a guitar string plucked at L/5 with pickup point L/13. Good electric guitars give the performer a choice of at least two pickup points: one close to the bridge (L/18, L/19) and another farther away. The pickup point close to the bridge amplifies a lot of modes, producing deep, bright sound, good for solos. The other pickup point favors lowest modes and is suitable for a background rhythm.

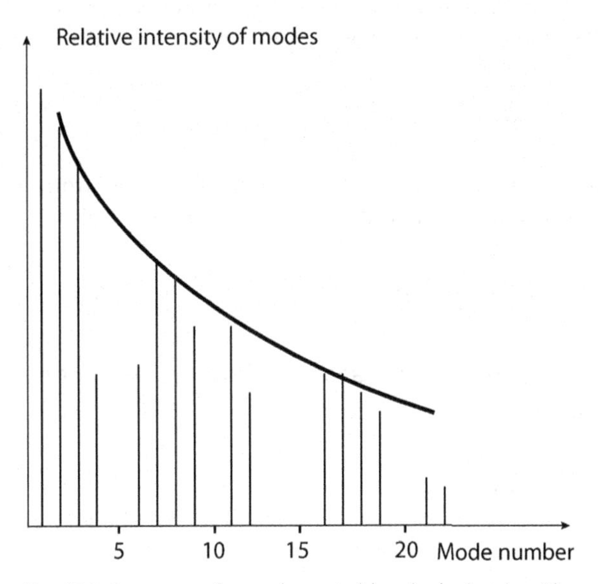

Fig. 11.6. Spectrum of sound created by plucked string. The plucking point is at position L/5, the pickup position is at L/13.

Example 11.7. What modes will disappear from a sound if we pluck string at L/4 and the pickup point is at L/11?

Answer:

All modes that have a node at the point of plucking or pickup will disappear from the sound. At L/4 there is a node for modes 4, 8, 12, 16, ...; for L/11 there is a node for modes 11, 22, 33, ...

11.2.3 Guitarmaker's Rule

The neck of a guitar has multiple frets—thresholds, which are approximately 2 mm high. On most modern western guitar-like instruments, frets are metal strips inserted into the fingerboard (Fig. 11.7). On historical instruments and some non-European instruments, pieces of string tied around the neck serve as frets. Frets divide the neck into fixed segments at intervals related to a needed temperament. On guitars frets usually correspond to the modern temperament system, having 12 semitones per octave.

Fig. 11.7. Frets of the guitar.

The frets on a guitar get closer and closer to each other while receding from the nut. Pressing a finger against the fret reduces the length of a string from a bridge to this fret and, as a result, increases the frequency of the sound. The ratio of frequencies between two frets corresponds to one semitone, or 1.059, the number which we already know from previous chapters.

$$\frac{f_{next\ fret}}{f_{first\ fret}} = 1.059 = \frac{distance\ to\ first\ fret}{distance\ to\ next\ fret}$$

That's it: the ratio of frequencies is opposite to the ratio of distances from corresponding fret to a bridge. Having this in mind, we can formulate the guitarmaker's rule:

$$distance\ to\ first\ fret - distance\ to\ next\ fret = 0.059L \cong \frac{L}{18}$$

Each next fret should be placed at distance L/18 from the previous one, where L is the length from previous fret.

This explains why frets get closer and closer to each other. Let the position of the first fret be at the distance 100 cm from the bridge. Then the second fret we should position at the distance 100 cm/18 = 6 cm from the first one. But the third fret we should place at the distance (100 cm – 6cm)/18 = 6.7 cm from the second one because the effective length of string is measured now from a second fret.

Example 11.8. The distance between bridge and first fret is 120 cm. At what positions should you place second, third, and fourth frets?

Answer:

We should position the second fret at 120cm/18 = 5.7 cm from the first. The third fret should be positioned at distance (120 cm – 5.7 cm)/18 = 5.3 cm from the second one. Finally, the fourth fret should be positioned at the distance (120 cm – 5.7 cm – 5.3 cm)/18 = 6.9 cm from the third one. As we can see, the distance between frets gradually decreases.

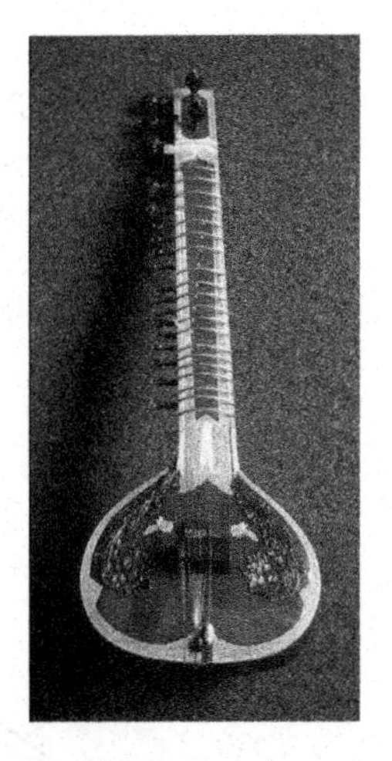

Fig. 11.8. Indian sitar: some frets are moveable.

Such positions of frets are completely defined by usage of western-European temperament. The Indian sitar, for example, can be tuned to different temperaments because some frets on its neck are movable (see Fig. 11.8).

11.3 The Piano

11.3.1 Some Words About the Mechanics of the Piano

We will not discuss in detail the extremely complicated construction of a piano which, moreover, depends on the particular model. Here is the information we should know:

1. The strings of a piano are not plucked, but **struck** with a little hammer after the player hits a particular key (Fig. 11.9). Hammers are covered with a soft felt. The quality of felt influences the character of the sound. The soft, cushy hammers make a gradual contact with strings, cutting the high-frequency modes, making the sound duller. Modern hammers are made so that they act quite softly if they hit the strings with small force; but faster, stronger, and more abrupt impacts bring into play the harder inner layers. As a result, soft and loud sounds on a piano can have not only different loudness but also different color.

2. Each string is dampened with a felt damper. These felt dampers are raised off the string when the key is hit and are pushed against the strings again when the key is released.

3. The piano has also a possibility to manage the decay times of the notes by using its **pedals**. Usually pianos have three pedals. The right one, often incorrectly called the “loud pedal,” is a **damper pedal**, which means the pedal that manages the system of

dampers. This pedal holds the entire set of dampers off the strings so a string can continue to vibrate even after its key is released, and other strings whose keys are not struck can resonate in sympathy. The middle pedal, which is often omitted for the upright models, is the sostenuto pedal, and it raises the dampers for whichever keys are down when the pedal goes down, but lets all other dampers operate normally. The left pedal, called **soft** pedal or **una corda** pedal, works differently on upright models

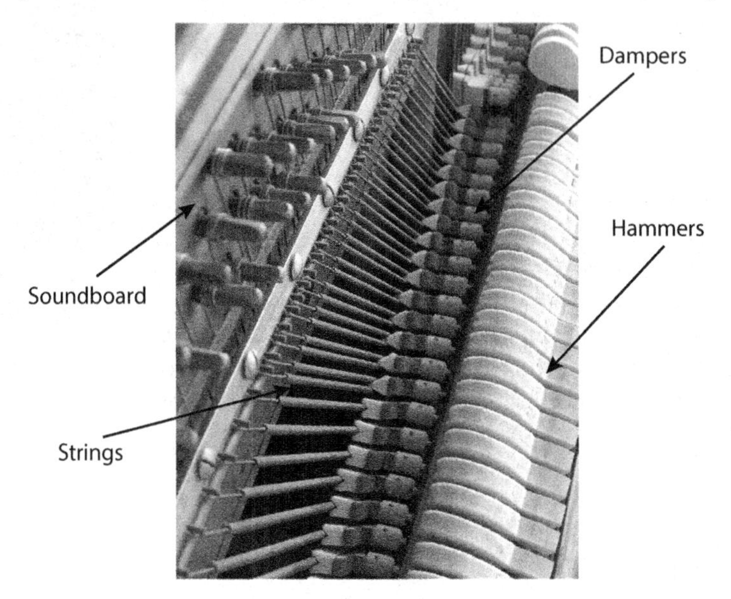

Fig. 11.9. Inside of a piano: strings, hammers, dampers, and part of the soundboard are shown.

and grand pianos. On upright models this pedal moves hammers closer to the strings so they do not hit the strings hard. On the grand piano this pedal shifts the entire mechanism to the right so that the hammers strike only one string out of two or three corresponding to each key.

11.3.2 Decay Rates of the Piano Tones

The piano is an instrument whose range covers practically all the musically important range of frequencies: starting from A_0 (26 Hz) up to C_8 (4186 Hz). This is definitely a great advantage of a piano as well as other keyboards but also produces some problems.

We know from a previous chapter that the contact time of a hammer or of a mallet has a crucial influence on the height of produced note. This means the contact of a hammer with a treble string of a piano must be extremely short.

Another problem is that the high-frequency vibrations die away more quickly. You can easily measure it by yourself on any available piano. The 60 dB decay time is 10–20 s for the extreme bass (depends on the particular instrument), 2–4 s for the midrange, and only half of a second and even less for the extreme treble. The initial transient component of sound is usually as equally bright for the full range, but the steady consequent sound of treble dies out really quickly.

The decay of piano tones is more complicated than exponential simplified decay, as discussed in a previous chapter. The tones begin to decay rapidly, but then the decay switches to a slower rate. In midrange, for instance, the initial decay rate is 2–3 s and changes after this to more than 10 s.

There could be several explanations why the decay tones of a piano behave in such sophisticated ways. One of them is that the string can vibrate not only up and down (for a grand piano) but also sideways. So, for each frequency there is, in reality, a couple of normal modes in vertical and horizontal directions. The hammer excites almost only vertical mode, but no one hammer is exactly vertically oriented. So, there is always a little addition of horizontal mode. The vertical mode dies away faster because it has efficient coupling with a bridge and soundboard; meanwhile, horizontal mode can exist longer because a soundboard responds poorly on it, and less energy is pumped from a string to a soundboard. Another possible explanation of non-exponential behavior of decay we will discuss in the subchapter "Multiple stringing."

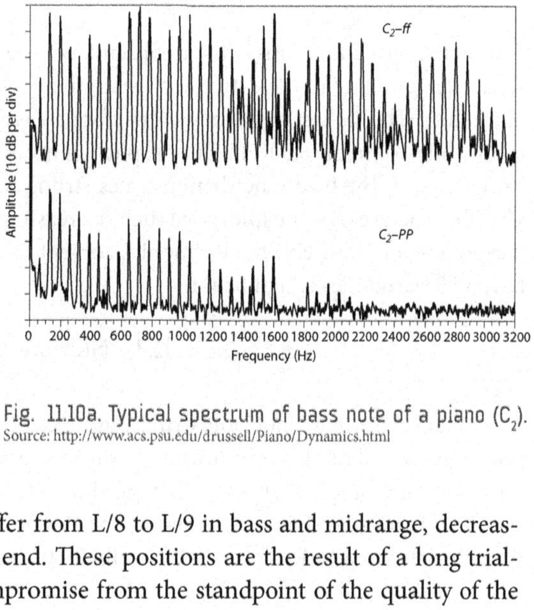

Fig. 11.10a. Typical spectrum of bass note of a piano (C_2).
Source: http://www.acs.psu.edu/drussell/Piano/Dynamics.html

11.3.3 The Typical Spectra of Piano Tones

The point of impact of a hammer, exactly like a plucking point on the guitar, will eliminate from the spectrum of the key all modes that have nodes at this point. The positions of a hammer in various pianos differ from L/8 to L/9 in bass and midrange, decreasing to as little as L/10 or L/12 at the treble end. These positions are the result of a long trial-and-error process and are a reasonable compromise from the standpoint of the quality of the sound. Elimination of the 10th or 12th harmonic after A_3 (220 Hz) does not influence the color of sound because the frequencies are already beyond the audible range.

The spectra of piano tones do not allow simple generalization as we were able to do for guitar strings. The reason is, again, the very broad range of frequencies. The examples of spectra of three different notes (bass, midrange, and treble) are shown in Figs. 11.10a–c. The duration of contact usually ranges from

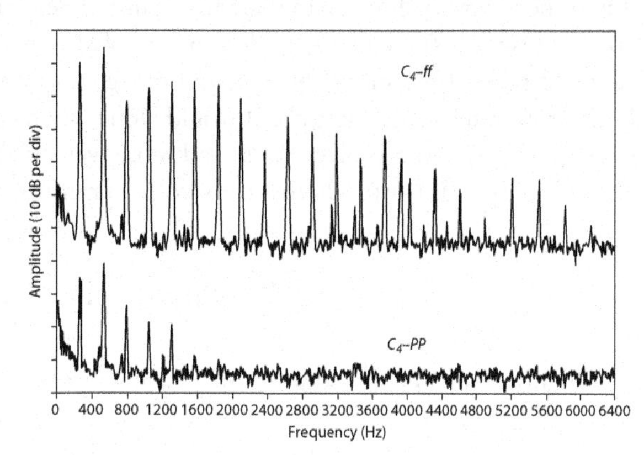

Fig. 11.10b. Typical spectrum of midrange note of a piano (C_4).
Source: http://www.acs.psu.edu/drussell/Piano/Dynamics.html

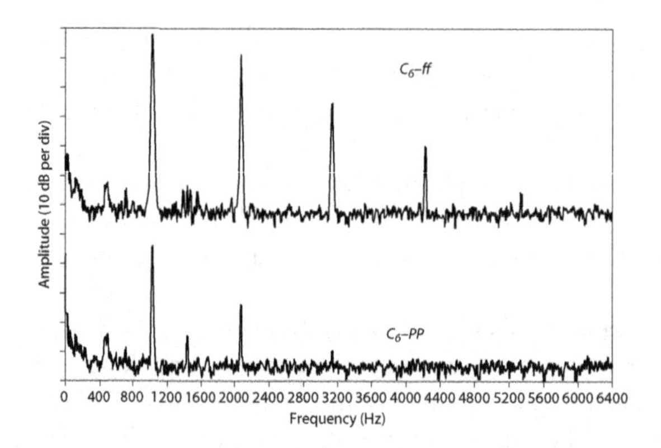

Fig. 11.10c. Typical spectrum of treble note of a piano (C_6).
Source: http://www.acs.psu.edu/drussell/Piano/Dynamics.html

5 ms in the bass and 0.5 ms at the treble end. For louder notes this contact is a little bit shorter; for softer, longer. Only in the extreme bass are the contact times are short enough not to kill upper harmonics. The highest frequency shown for bass is approximately 1.5 Khz, for midrange 4.0 KHz, meanwhile, for treble 6.3 KHz. But the pitch of notes themselves changed by 4 octaves, or 16 times in frequency! The bass tone demonstrates strong harmonics, decreasing, at least for first 12, as 12 dB per octave. For the midrange such decreasing of harmonics in amplitude lasts only up to 6 harmonics, and for treble the harmonics content falls rapidly, at 22 dB per octave so that nothing beyond 5 harmonics is significant.

11.3.4 Multiple Stringing

Most piano keys have not one string but a set of two or three strings. And these strings are purposely tuned not to be in unison. Let us discuss what would happen if, say, three strings on one bridge are tuned in unison? They produce a force against the bridge three times bigger than the force from one string. As a result, the bridge and the soundboard will vibrate with three times bigger amplitude. This results in efficient energy transfer and fast decay.

But suppose now that our strings are tuned not in perfect unison, but slightly off 1/3–1/2 of one Hz, for instance. The human ear does not resolve so small an "off." But approximately after 1 s of sound these three strings become out of step and cannot act on the bridge in cooperation. Each string radiates its energy as it would alone, giving a slower rate of decay. Moreover, slight mistuning of the strings gives the sound more warmth, and the slow beats between the strings (1/3, 1/2 of one Hz) also help make the voice of our instrument more sophisticated.

11.3.5 Inharmonicity of the Piano

We already know that the fundamental frequency of a string can be written as

$$f_n = \left(\frac{1}{2L}\right)\sqrt{\frac{F_t}{\left(m/L\right)}}$$

It would seem at first that we have a lot of freedom to construct a piano of any size. But alas, no.

1. The diameter of a string (either small or large) creates problems. The very light string is not able to set into motion a massive soundboard, so the resulting sound is weak. The thick wires, on the other hand, are not flexible enough. This stiffness provides extra restoring force, which, acting together with tension, shifts the resulting frequency up. But the problem is that the shift of higher modes is bigger than the shift of the fundamental, and this means the spectrum does not demonstrate harmonic series anymore. This effect is called **inharmonicity**. This stiffness makes the string perform like a rod, and the result is a sound that feels tinny. The typical spectrum of the well-tuned uppermost piano looks like f_1, $2.01f_1$, $3.03f_1$. The inharmonicity is very small, but audible.

   Inharmonicity largely affects the lowest and highest notes in the piano, and is one of the limits on the total range of a piano. The lowest strings, which would have to be the longest, are most limited by the size of the piano. The designer of a short piano is forced to use thick strings to increase mass density and is thus driven into inharmonicity. The highest strings have to be under the greatest tension, yet they must also be thin to allow for a low mass density. The limited strength of steel forces the piano designer to use very short strings, which leads again to inharmonicity.

   Inharmonicity also produces problems with the tuning of a piano. In order to tune an octave, a piano tuner must reduce the rate of beats between the second harmonic of a lower note and a higher note until it disappears. But because of inharmonicity, this second harmonic will be sharper than a note one octave higher (the ratio of, say, 2.02/1 to a lower note instead of the ratio of 2/1 as for a perfect octave), making either the lower note flatter or the higher note sharper, depending on which one is being tuned to. To produce an even tuning, the skilled tuners begin by tuning an octave in the middle of the piano first, and then proceed to tune outwards from there; notes from the upper range are not compared to notes in the lower range for the purposes of tuning. Pretty often in the top octave, where strings are relatively stiff, the octaves, such as C_7–$C_{8,}$ are "stretched" to the ratio 2.025/1.

2. If the string is too short, some tone quality will be sacrificed. This is obvious: the short string is dangerously close to just a metallic bar (as we considered in a previous chapter), and metallic bars do not have any harmonic spectra. Still, the length can vary over quite a wide range, from tens of centimeters to a couple of meters.

3. The string's tension cannot exceed, or even be close to, the breaking strength of the string. This depends on the composition and thickness of the string. There is no strict bottom limit for tension, but lesser tension drastically reduces the ability of a string to transfer energy to the soundboard. Just imagine a piece of rope hanging loosely between two supports.

How can we minimize the difficulties described above? The answeres were developed over many years by skilled technicians, and include:

1. Multiple stringing in the treble will help to provide more energy to the soundboard.
2. Instead of longer, the strings for bass should be heavier.
3. For treble, use very short strings rather than very thin and light ones. This process of shortening a string should be also be accompanied by reducing the diameter to prevent too much stiffness.
4. Set the force of tension on all strings close to the maximum they can tolerate. This will provide a more vivid, enduring tone.

Summary, Terms, and Relations

The general formula for a frequency of a normal mode number *n* on a thin string:

$$\lambda_n = \frac{\lambda_1}{n};$$

$$f_n = \frac{V}{\lambda_n} = n\frac{V}{2L} = nf_1$$

where n is any integer and

$$\lambda_1 = 2L \text{ and } f_1 = \frac{V}{\lambda_1} = \frac{V}{2L}$$

are wavelength and frequency of the fundamental mode of a string.

The modes that have nodes at points of plucking and pickup disappear from the recipe of the resulting sound.

Guitarmaker's Rule: Each next fret should be placed at distance L/18 from the previous one where L is the length from previous fret.

Effect of inharmonicity of a piano: when the string demonstrates the spectrum slightly off the harmonic series.

Questions and Exercises

1. If the velocity of transverse waves on a certain string of length 0.5 m is 180 m/s, what is the frequency of its fundamental mode? Of its second harmonic? What are the wavelengths of the first three modes of this string?

2. If you double the tension applied to a string, by how much is its fundamental frequency increased?

3. If a string of mass 5 g and length 1 m is placed under tension of 200 N, what is the speed of the transverse wave on it?

4. What are the wavelengths and frequencies of the first three normal modes on the string in the previous example?

6. Where would you pluck a guitar string to most effectively excite mode 3? Where would you touch it with your finger to kill modes 1 and 2 but leave mode 3 vibrating?

5. What are the positions of the nodes and antinodes for mode 6 on a thin string?

7. Discuss a resulting spectrum of a guitar string if the plucking point is a L/4 and pickup point at L/11.

8. What notes of a piano keyboard demonstrate the longest decay?

9. Discuss why, for two strings of the same diameter under the same tension, the shorter one shows more inharmonicity in its vibrations.

10. Suppose you pluck a guitar string that sounds the note G_2, and you touch your finger lightly exactly at L/4 from the nut. What pitch will you hear now?

Chapter 12

The Violin

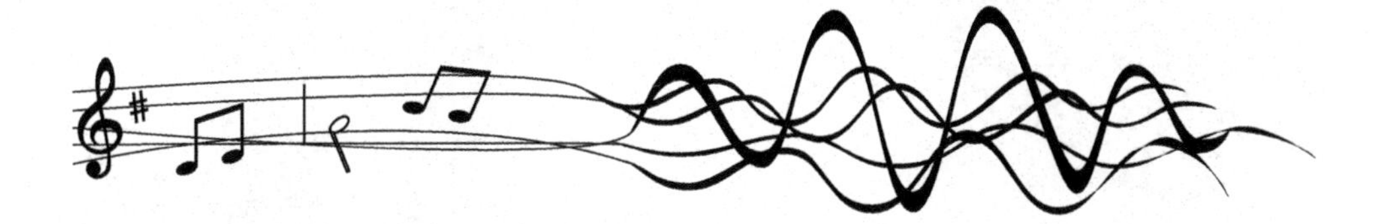

12.1 Construction of the Violin

The violin family, the heart of any chamber or symphonic orchestra, has a long and interesting history. The first precedent of the violin (lyre) appeared in ancient Greece or maybe even earlier—we don't know because the history of the development of violins is not documented well enough. Practically any culture has in its "orchestra" a string instrument that can be considered as some form of violin. Fig. 12.1 shows the violins of different cultures: they have a different number of strings and a different form of the bow and body, but they may be called violins because they use bowed strings. The oldest documented European violin having four strings like our modern ones was constructed around 1555 by the Italian master Andrea Amati. But even this date is doubtful. All that we know is that the oldest date inside a violin, with the name of Charles IX, was created c. 1560. Three great families, who lived in Cremona, Italy in the sixteenth and seventeenth centuries, brought us the violin in the modern shape: Amati, Guarneri, and Stradivari. The long-lasting secret of the deep, velvety voice of their instruments disturbed a lot of generations.

Some changes occurred in the construction of the violin in the 18th century, particularly in the length and angle of the neck, as well as the more massive bass bar. The old instruments have undergone these modifications—and hence are in a significantly different state than when they left the hands of their masters—which influenced their voice. But these instruments in their present condition set the standard for perfection in violin craftsmanship and sound, and violin makers all over the world try to come as close to this ideal as possible.

The main features of violin construction are shown in Fig. 12.2. Note the extremely delicate **neck** made of maple. Being so delicate, it withstands the tension of four strings and supports

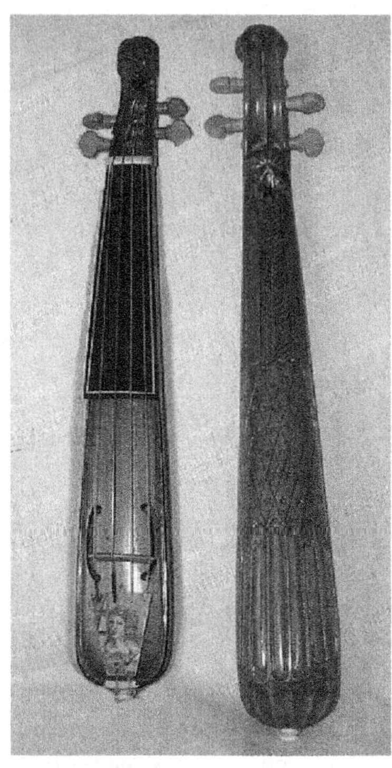

Fig. 12.1. Different types of violins, instruments using the bowed strings.

the fingerboard, made of ebony. The playing tension of each string varies from 40 to 90 N.

In the middle of the body of the violin, a **bridge** supports the strings. The violin has a finer, less rigid, and relatively higher bridge (Fig. 12.3b) than the plain rectangle, stiffly mounted on the top plate, that we see in old bowed instruments. The older type of rectangle bridge (Fig. 12.3a) practically does not allow the horizontal motion of strings; only vertical motion of strings is favored. The more modern bridge has a somewhat uneven shape, which enables it to move, pushing in and pulling out on the top plate. As a result, the plate accepts more of the string's energy.

The strength of the plates is intentionally reduced down the entire length of the edge that is attached to the ribs; this allows for easy vibration. An inset narrow decorative edge, called **purfling**, runs around the edge of the spruce top and provides some protection

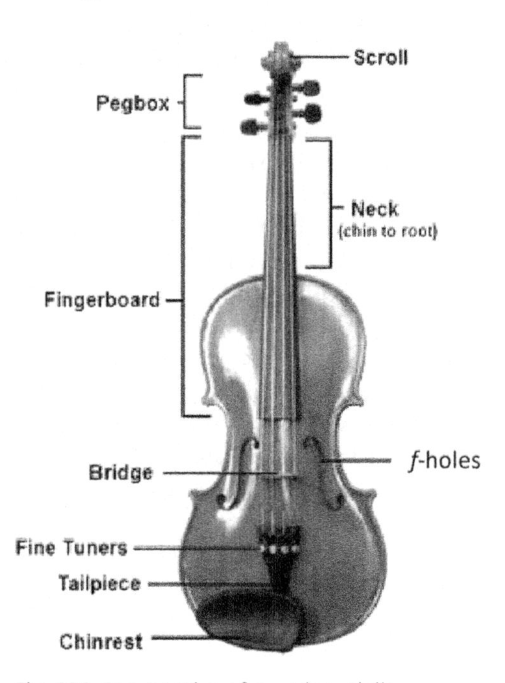

Fig. 12.2. Construction of a modern violin.

against cracks originating at the edge. It also allows the top to flex more independently of the rib structure. Inexpensive violins may have no purfling at all.

Perhaps the most sophisticated aspect of the violin's composition is the structure of the **body**. The plates (both top and back) are sculpted precisely in order to achieve the proper curvature and thickness, which can be as little as 0.2 cm at the edges. The wood's grain and the strings run parallel to each other; since the plate is relatively supple across the wood's grain, the parallel placement makes it stiffer in this direction.

Curly maple is typically used for the back plate, while spruce, the famous "musical" wood (which is used also for soundboards of guitars and pianos) is the wood of choice for the top plate. Here, a thin shield of varnish is applied. The most important things about the varnish are that it should be applied in an exceptionally thin layer and should have good elastic properties. A thick coat of varnish will pretty soon create cracks, damaging the instrument's voice.

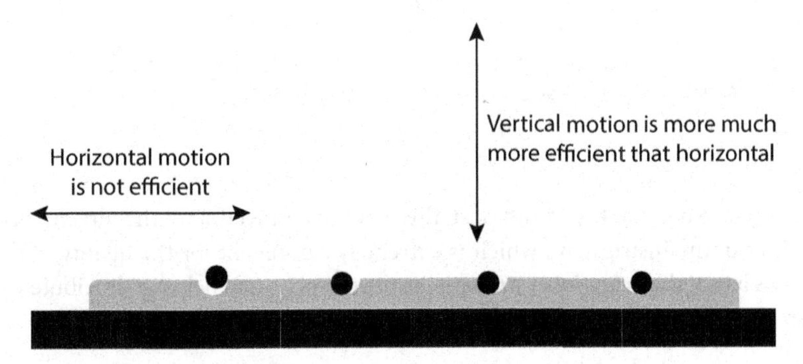

Fig. 12.3a. Example of a flat bridge: the horizontal motion of strings is effectively suppressed.

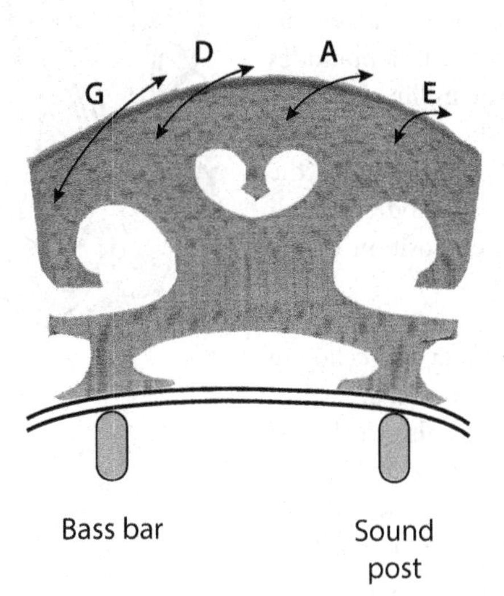

Fig. 12.3b. Bridge of a modern violin: both horizontal and vertical motions of strings are favored. The treble foot is relatively fixed over the sound post. It acts as a pivot for the bridge motion.

There are long-lasting legends, invented by old masters, about the secret recipe of varnish. Definitely, the right varnish serves the purpose of the proper aging of the instrument, but the quality of the instrument is obvious before the varnish is applied.

The total tension in the four strings of the violin is typically around 250 N. This results in a downward force exerted on the bridge of approximately 100 N (the weight of 22 lbs), called the down-bearing (Fig. 12.4). The **sound post** (also called soul post) provides a connection between the top and back plates, so the back plate helps to bear the steady load of down-bearing. The sound post also supports the treble foot of the bridge, like the fulcrum of a rocking lever, when

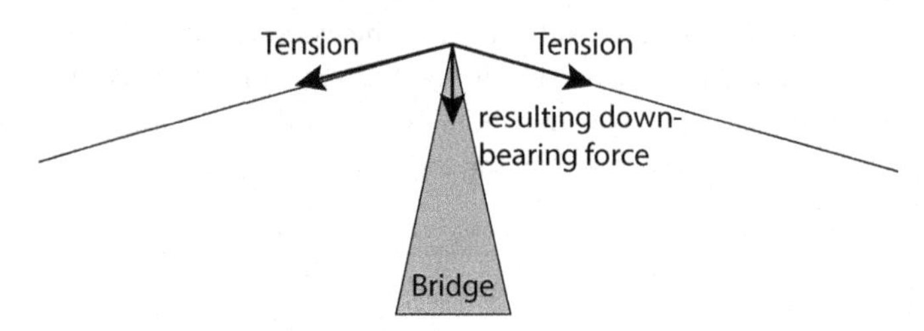

Fig. 12.4. Forces exerted on the bridge of the violin by a string. The forces of tension result in the down-bearing force.

the strings vibrate. Note that the sound post, the point of connection of the top and back plates, is off the center of the instrument, which is extremely significant for the quality of the violin's voice. The **bass bar** of the bridge also strengthens the top plate and helps to distribute vibrational energy from the bridge over the top plate.

Just like any other object, the violin's body has many normal modes of vibration. The modern optical techniques make it possible to visualize these vibrational motions with amplitudes less than a thousandth of a millimeter. This has been used to map out the shapes of some of these normal modes (Fig. 12.5). Note the asymmetrical shape of these modes, which is the result of the off-center position of the sound post.

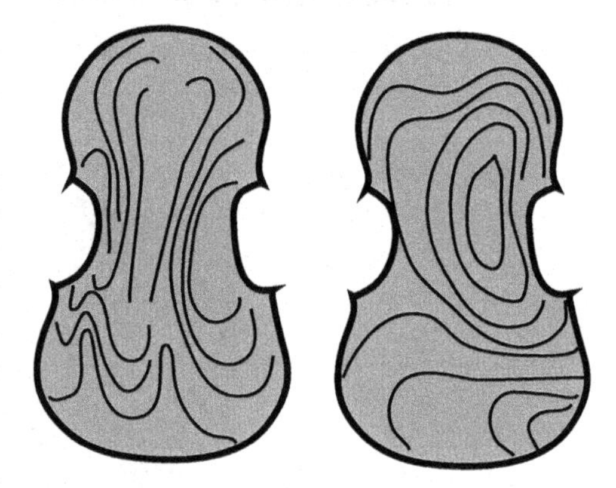

Fig. 12.5. Normal (natural) modes of the body of the violin. Note that modes are not exactly symmetrical relative to the center of the instrument because of the off-center position of the sound post.

The general features of the viola and cello are similar, but each has some unique character of its own. The huge difference is also in the direction of the sound radiation from the instrument, which will be discussed further later in this chapter. The sizes of these instruments are not in direct proportion to the wavelengths of the produced sound. The cello range begins at C_2, the viola at C_3, and the violin at G_3. However, violas are not 50% but only 15–20% bigger than violins. Cellos are a little more than twice a violin's size; if the sizes were directly proportional, the cello would have to be three times as big as a violin. This non-proportionality of sizes leads, for instance, to the pretty weak low register of violas.

12.2 Strings and the Bow

The voice of a violin is created by the motion of the bow against the strings. A violin is usually played using a bow consisting of a stick with a ribbon of horsehair strung between the tip and frog (or nut, or heel) at opposite ends. A typical violin bow may be 75 cm overall and weigh about 60 g. The hair of the bow traditionally comes from the tail of a grey male horse (which has predominantly white hair) though some cheaper bows use synthetic fiber. The bow is rubbed occasionally with rosin to make the horse hair stickier. The reason for using horse hair lies in the perfect cylindrical shape of this hair and occasional "hooks"—not too close to each other, not too rare (Fig. 12.6).

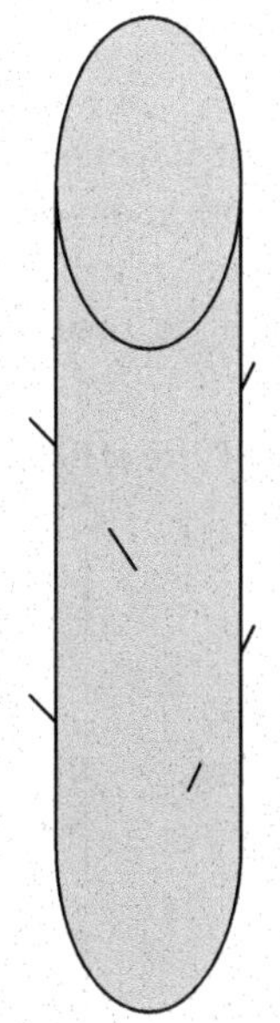

Fig. 12.6. Sketch of a horse hair: perfectly cylindrical in form, having just enough hooks to provide proper sticking with the strings.

Strings were first made of sheep gut (commonly known as catgut), or simply gut, which was stretched, dried, and twisted. In the early years of the twentieth century, strings were made of gut, silk, aluminum, or steel. Modern strings may be gut, solid steel, stranded steel, or various synthetic materials, wound with various metals, and sometimes plated with silver. Most E strings are unwound steel, either plain or gold-plated. Currently, violin strings are generally not made of gut, with the exception of violin strings used to play music from the Renaissance, Baroque, or early Classical periods.

It is interesting to mention that the word catgut has nothing to do with cats at all. Catgut is a type of cord that is prepared from the natural fiber found in the walls of animal intestines. Usually sheep or goat intestines are used, and the word "catgut" may be just an abbreviation of the word cattlegut, which folklore turned into the entrails of the cat.

12.3 The Mechanism of the Bowing: Stick and Slip

One of the most important questions we should answer while discussing the violin is how the steady motion of the bow causes the string vibration. Why is it that the string cannot stay in one position, slightly aside from equilibrium, while the bow slides relative to it? If you are not an experienced player, or, better to say, if you try to play violin for the first time in your life, you will not push the bow hard enough against the strings but will allow it just to skitter across the strings. But this will create a very small vibration.

To understand this mechanism, we first should consider the very important force that plays a dramatic role in this process: the force of friction.

12.3.1 Force of Friction

There are two types of friction. The first one, static friction, acts when we try to move one object relative to another, but our force is not big enough to overcome the mutual friction of the two surfaces. Anyone who has tried to move a refrigerator knows the difficult nature of friction. The dependence of the force of friction in an applied effort is shown in Fig. 12.7. We apply a force of

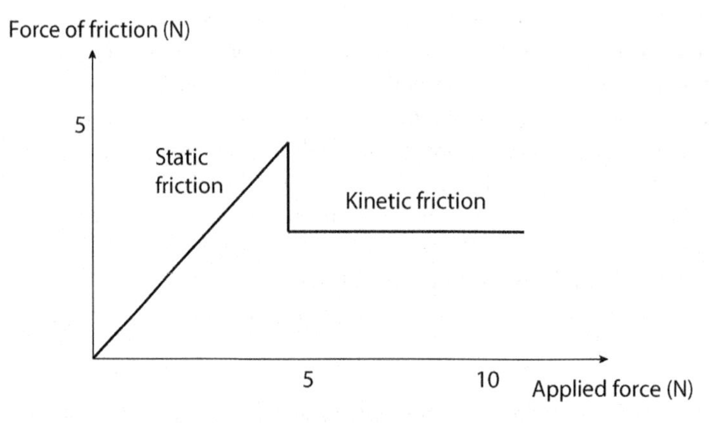

Fig. 12.7. Dependence of force of friction on the applied force. If there is no relative motion (static friction), the force of friction increases linearly. After reaching its maximum value for surfaces in contact, the force of friction drops and afterwards stays the same (kinetic friction).

1 N (all numbers are just for example), and the force of friction that matches this 1 N; the object does not move. We increase our efforts to 3 N, but the force of friction also increases, keeping

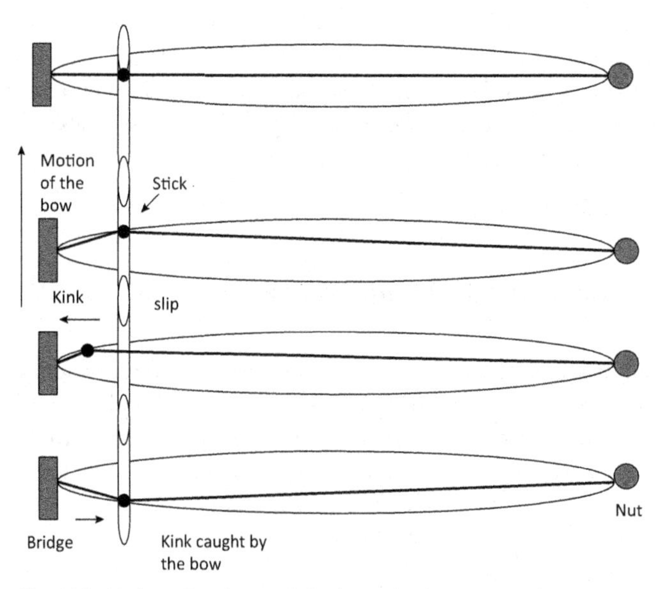

Fig. 12.8. Motion of string and the bow: the bow moves in one direction (up), the string sticks and slips many times. The motion of a kink towards the bridge is shown.

our object at rest. But sooner or later, our efforts become so strong that the two objects start to move relative to each other. At the very last point that they are still at rest we are given the maximum possible static friction between two particular surfaces or objects. Also, anyone who has tried to move a refrigerator knows that after this heavy thing starts to move, it looks like everything became easier: our refrigerator literally slides with much less effort than we applied to budge it. This can be seen again in Fig. 12.7: in comparison to maximum static friction, the force of friction between two sliding surfaces (called kinetic friction) is much less.

12.3.2 Stick and Slip Mechanism

But what does the moving of a refrigerator have to do with the bowing mechanism of the violin? Let us check Fig. 12.8. In the very beginning of the bowing, the sticky horse hair of the bow carries the string aside. There are several forces that struggle with each other: two forces of tension, directed along the string, give the total restoring force, which tries to bring the string back; and the force of friction, which for a while keeps the string stuck to a horse hair. Sooner or later, the force of friction can no longer match the restoring force of total tension, and, as a result, the string starts to slide relative to the bow. The force of kinetic friction is much less and cannot catch the string while it is moving. The string misses the equilibrium point and reaches the opposite point of maximum displacement, and on the way back, at the moment when the

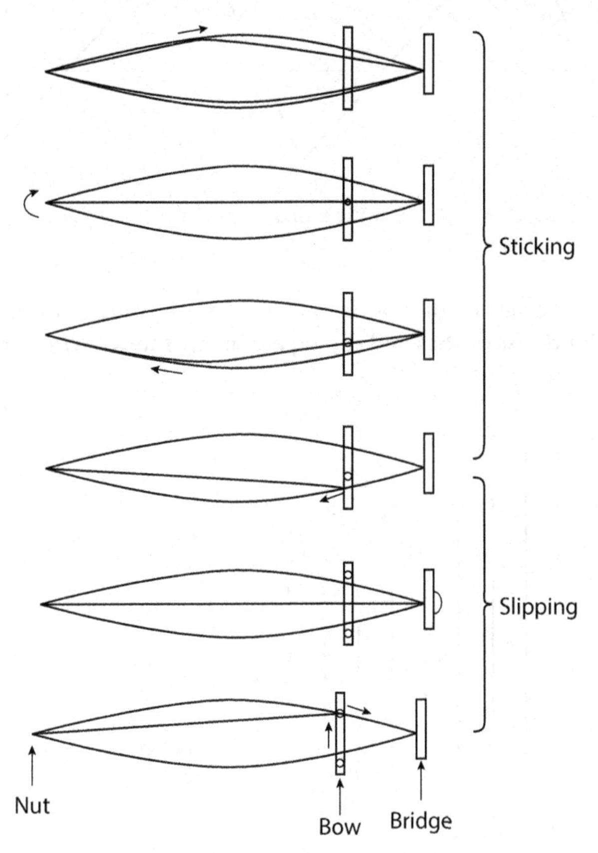

Fig. 12.9. Stick and slip mechanism mode in detail.
Source: Musical Acoustics by Donald Hall, 2002 Cengage Learning, Inc.

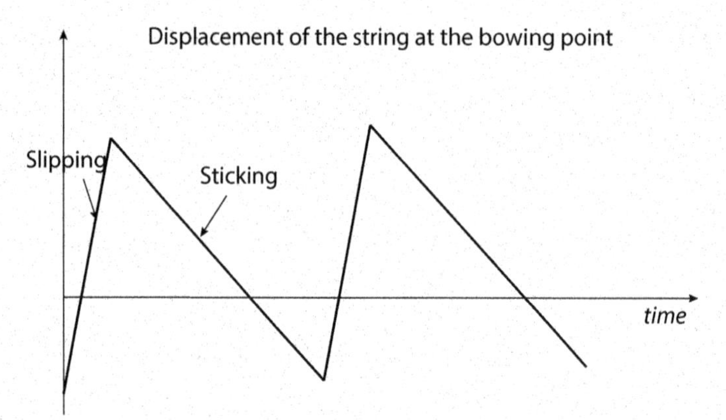

Fig. 12.10a. Displacement of string while playing close to the bridge.

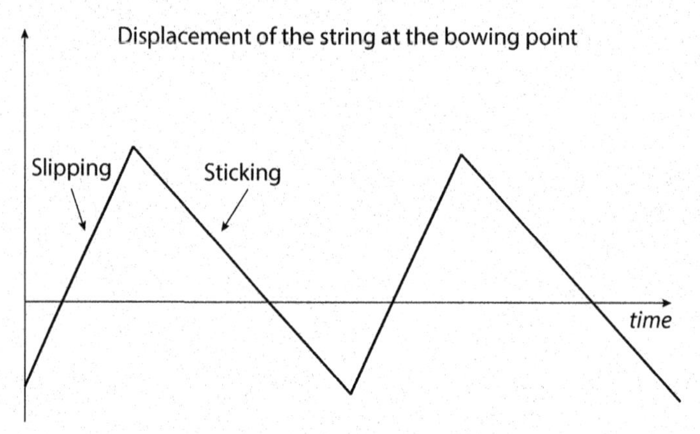

Fig. 12.10b. Displacement of the string while playing at a bigger distance from the bridge.

relative velocity of the string and the bow is zero, the string is "caught" by the bow to be carried again. Note **that this process of stick and slip repeats many times during one long bow stroke.**

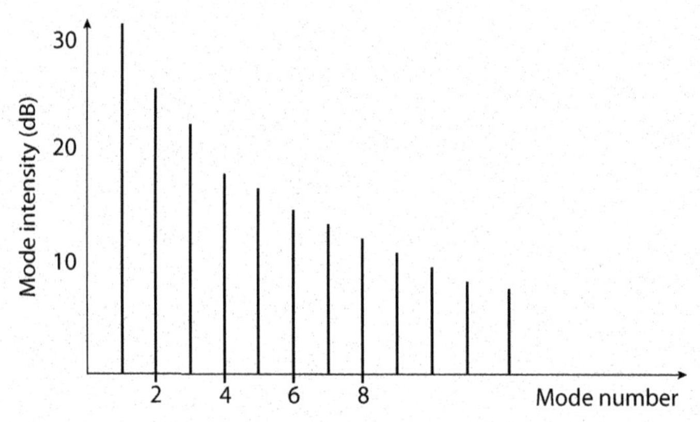

Fig. 12.11. The idealized spectrum of a string with bowing point close to the bridge.

Still, it is not obvious what are the best and perfectly smooth ways to use this mechanism. If we think about the motion of a string as perfectly sinusoidal, it is hard to understand how sticking and slipping would happen at the same points and maintain the steady vibration rather than just a random noise. Better to explain it in terms of a traveling kink. The sudden release of a string sends a little kink traveling along the string. That kink is always moving toward the bridge. There is a pretty simple reason for it: the bow is always closer to the bridge than to the nut, so the total force of tension is not exactly vertically down but a little bit toward the bridge. That kink is reflected from the bridge, and when it arrives back at the bowing point, it gives that part of the string a little kick that ensures that it sticks to the bow again. And after reflecting from the nut, completing its round trip, this kink helps break the string loose at just the right time to start another cycle. When this self-regulating mechanism is working properly, the kink moves on the string, as shown in Fig. 12.9.

The sideways displacement of the string as a function of time is shown in Fig. 12.10 for two positions of the bowing point to the bridge: very close to a bridge (a) and at larger distance (b). The closer the bowing point is to the bridge, the shorter is the fraction of each cycle, during which the slipping occurs. It means also that the form of a kink becomes closer to a form of a sawtooth wave. Meanwhile, a point far from a bridge creates a rather triangular kink because slipping and sticking takes approximately the same time. As we recall from Chapter 9, the triangular wave has pretty much suppressed even harmonics, which makes the voice of the instrument poorer.

The idealized spectrum corresponding to a bowing point close to the bridge is shown in Fig. 12.11. The amplitude of harmonics decreased 6 dB/octave.

12.3.3 The Bowing Force and Performance

The bowing force is a downward force that holds the bow against the strings. Note this is not a frictional force, which is directed parallel to the bow. But friction depends on this downward

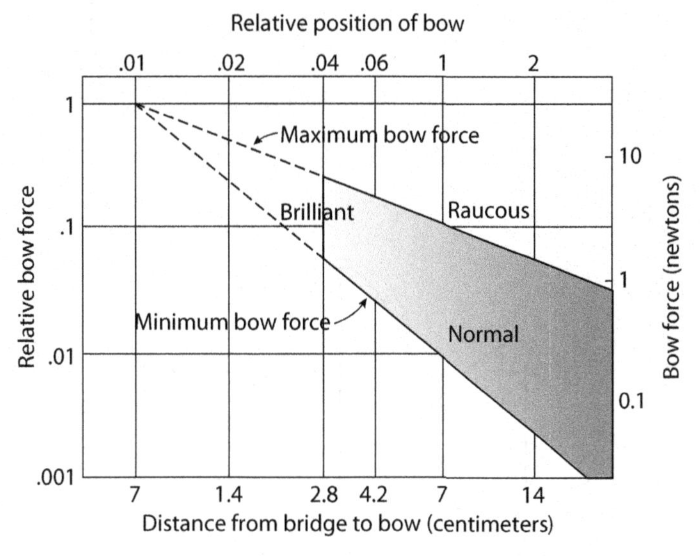

Fig. 12.12. Values of bowing force providing proper playing as function on distance from a bridge at given speed of bowing. Note that the corridor of proper bowing forces is getting narrower when the bowing point approaches the bridge.
Source: Musical Acoustics by Donald Hall, 2002 Cengage Learning, Inc., after Schelleng (1974)

force; to be exact, it is proportional to it. If the bowing force is too small, the slipping occurs too soon, before the kink returns to its initial point. This can set into motion two or even more traveling kinks, creating the oscillation of the string at one of the highest normal frequencies. This effect can be reached deliberately and more reliably by placing a finger on the node of the desirable mode, allowing it to vibrate undisturbed and damping away all lower modes (flageolet).

There is also an upper limit on the bowing force. If you push too hard, the kink will be unable to break the string loose from the bow at the right moment. The regularity of vibrations in this situation is no longer self-supported, and the sound gets an erratic, squawky character.

Fig. 12.12 shows the corridor inside of which the bowing force should be maintained to get normal playing (from Schelleng, 1974). This range becomes narrower as the bowing point is moved closer to a bridge. Beginners play well far from the bridge where the wider range of allowed force is more forgiving for their irregularities. Professionals who have much more control of the instrument play closer to a bridge, achieving greater volume and brilliance, partially because the spectrum comes closer to a sawtooth wave (Fig. 12.11), and partially because the bigger downward force causes more friction, delivering more energy to the strings.

The bowing speed also influences the sound. For each bowing speed a diagram similar to Fig. 12.12 could be constructed. In summary, louder playing requires greater bowing speed or

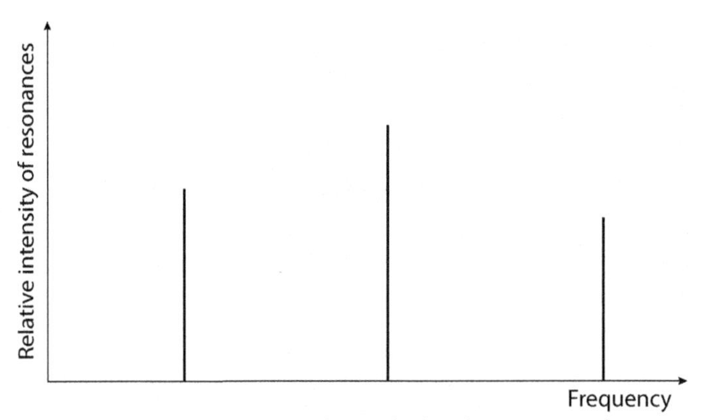

Fig. 12.13a. Strong resonance peaks. Such situation corresponds to woodwind instruments.

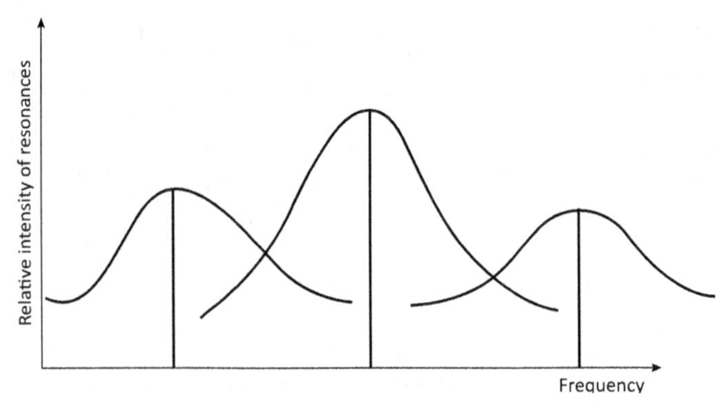

Fig. 12.13b. The resonating peaks overlap, what corresponds to soundboards of keyboards.

bowing closer to the bridge. This in turn requires greater bowing force in the narrow range to maintain the proper oscillations.

12.4 Resonance of the Violin

We already have been introduced to the term "resonance." Resonance is the increasing of the amplitude of vibration when the frequency of the external force is close to the normal frequency of a system.

So far we considered this phenomenon for systems like masses on springs, which have only one degree of freedom. But the concept of the resonance is applicable also for sophisticated bodies with many degrees of freedom. Each natural mode now has its own resonant response, peaked at its own characteristic frequency (Fig. 12.13). Case (a) in that figure corresponds to strong resonances when we want only several well-defined pitches with nothing in between. This situation corresponds to woodwind instruments. Case (b) is what is required from the soundboard of all keyboards: the resonating peaks overlap so that all frequencies are equally enhanced.

The violin, although it can have some special, individual features to its peaks, should respond fairly uniformly for all notes. To understand what kind of main resonances a violin has, let us check loudness curves, as shown in Fig. 12.14 (taken from C. Hutchins "The Physics of Music"). Technically speaking, these are curves of SIL because what is measured here are decibels, not sones or phons. To create these curves the violin is played as loudly as possible on each note, and its SIL is measured with a sound level meter. This analysis gives us the voice of not just strings alone but also the resulting spectrum of the sound of the violin that already reaches our ears.

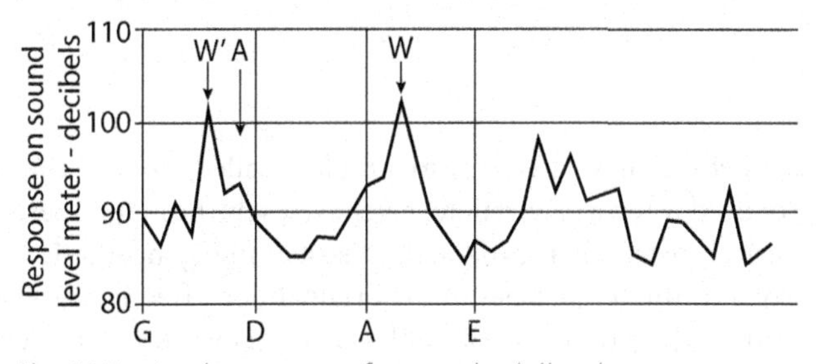

Fig. 12.14a. Loudness curve of a generic violin: air resonances are very weak.

Source: "The Physics of Violins" by C M Hutchins, 1962 Scientific American, Inc.

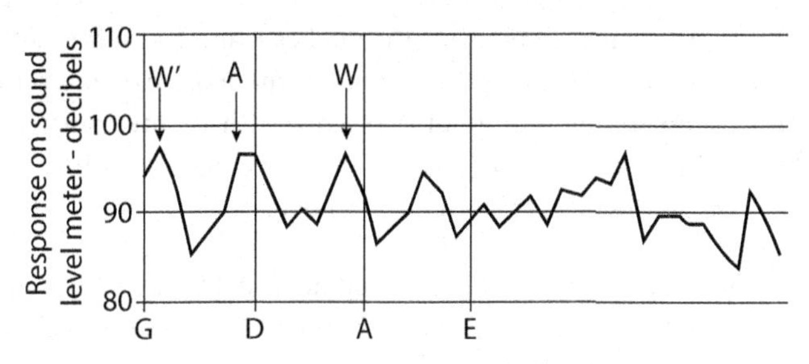

Fig. 12.14b. Loudness curve of Stradivari violin: air and wood resonances are almost of the same height.

Source: "The Physics of Violins" by C M Hutchins, 1962 Scientific American, Inc.

In Fig. 12.14 we see loudness curves for two instruments of very different quality. The (a) case is a violin with weak response between D_4 and A_4 but two huge maxima for Bs. The second curve demonstrates the response of a good Stradivari. Three resonances have practically the same height. Such form of resonances is crucial in determining of the quality of the instrument.

The open circle above each curve indicates the main **air resonance**, which is defined primarily by the size of the f-holes and the volume of a body rather than any properties of the wood. The f-holes work here as the opening of the Helmholtz resonator—air moves in and out through them. If the main air resonance is tuned to occur near the second open string (D_4 for violins), it creates a strong, solid sound for that note and other notes nearby. Higher cavity modes create little motion through the f-holes and are acoustically less important.

The solid dot to the right at each curve identifies the **wood resonance**, which corresponds to the lowest natural mode of the body and is completely defined by the mass and stiffness of the wood. It seems that the best position of it is near the frequency of the third string (for violin A_4). The poor quality of violin (a) is mostly because of the weakness of the air resonance and too great a distance between the air and wood resonances.

The solid dot to the right above both curves also represents a wood resonance, which is called **wood prime**. It is lower than the main resonance, about which we said that it corresponds to the lowest mode of the body. But if you check the curve, you'll see that the wood prime is one octave below the main wood resonance. Wood prime represents the notes whose radiated fundamental is extremely weak but whose second harmonics is strongly reinforced by the main wood resonance. So, the violin (b) produces its lowest notes with good strength and fullness. Violin (a) is weak on its three lowest notes because its main wood resonance is too high.

12.5 Wolf Tones

String instruments sometimes have a very nasty problem called the wolf tone. A wolf tone, or simply a "wolf," is produced when a played note matches closely the natural resonating frequency of the body of a musical instrument, producing a sustaining sympathetic artificial overtone that amplifies and expands the frequencies of the original note, frequently accompanied by an oscillating beating. This beating occurs because the artificial overtone is not exactly one octave higher than the basic note. The sound that is produced as a result of such sharp resonance is unsteady and reminds us of the howl of the animal.

Frequently, the wolf is present on or in between the pitches E and F# on the cello and around G# on the double bass. A wolf can be reduced or eliminated with a piece of equipment called a **wolf tone eliminator**. This is a metal tube and mounting screw with an interior rubber sleeve that fits around the offending string below the bridge. Different placements of this tube along the string influence or eliminate the frequency at which the wolf occurs.

An older device on cellos was a fifth string that could be tuned to the wolf frequency; fingering an octave above or below also attenuates the effect somewhat as does the trick of squeezing with the knees.

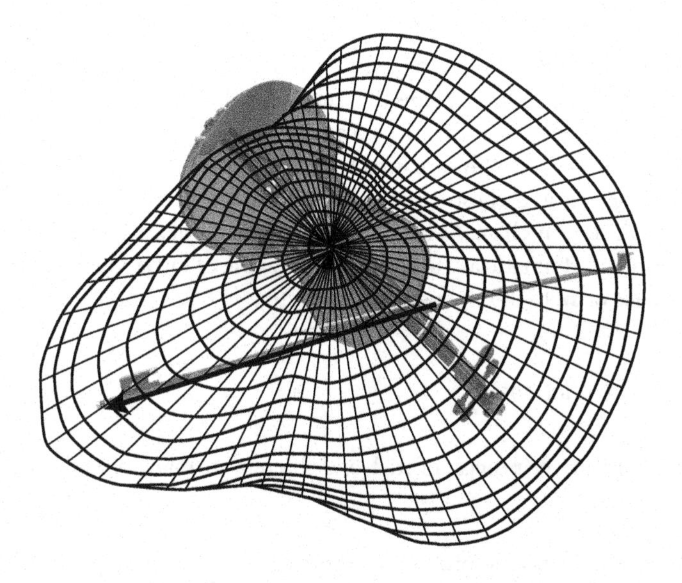

Fig. 12.15. Diagram on sound radiation from a violin. Direction on the most active radiation is shown.

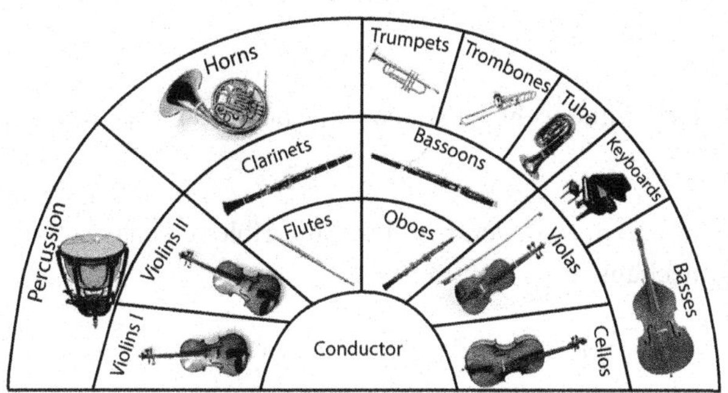

Fig. 12.16. Seating arrangements in symphonic orchestras. All strings, woodwinds and keyboards are arranged to provide the best sound radiation towards the audience.

It is interesting to mention one tale about a wolf. Historian Henry Shoemaker recorded that a hunter named Lewis Dorman from the Seven Mountains region of Pennsylvania would draw wolves out of the forests by playing wolf tones on an old violin. Shoemaker notes that Dorman was an accomplished violinist and frequently played by night, but the wolves would ignore his music and respond only when a wolf tone was produced.

12.6 Radiation of String Instruments

The bowed string instrument does not radiate equally in all directions. Moreover, the uniformity of this radiation strongly depends on the frequency range of played notes. Low frequency (200–500 Hz) has practically spherical distribution around the performer, but the higher the frequency gets, the bigger the amount of sound that is radiated from the top, the most active plate of the violin (Fig. 12.15). This explains the well-defined seating arrangements in symphonic orchestras (Fig. 12.16). The violins are always placed to the left of the conductor because in such position, the top plates of the violins are directed toward the audience. The second violins can sit to the right, like American orchestra seating, or they may sit to the left behind the first violins as required in the European chart. Cellos, violas, and basses also have a directed diagram of radiation, but it is not as dramatic as violins mostly because of lower voices. For instance, cellos can face the audience or sit to the right of the conductor.

Summary, Terms, and Relations

Stick and slip mechanism: the process of stick and slip repeats many times during one long bow stroke.

A good violin has air and wood resonances of approximately the same strength.

The violin radiates sound mostly from the top plate.

Questions and Exercises

1. If a violin A_4 string of length 40 cm is to be under the tension 60 N, what linear mass density should it have?

2. When you move your fingernail against a window, you may hear a terrible screechy sound. Explain this in terms of the bowing mechanism.

3. Discuss the principle difference between spectra of good and generic violins.

4. What is the difference in spectra when a performer moves a bow far from the bridge and close to the bridge?

6. Discuss the motion of a string during the long bow stroke. In what direction does the kink move? Sketch the motion of the string.

5. Discuss what the advantages are of having the sound post positioned off-center on the violin's body.

7. Why do the violinists always sit to the left of the conductor in big orchestras?

8. Discuss what differences it should make to a violin's tone if it had no f-holes.

Chapter 13
Flutes and Recorders

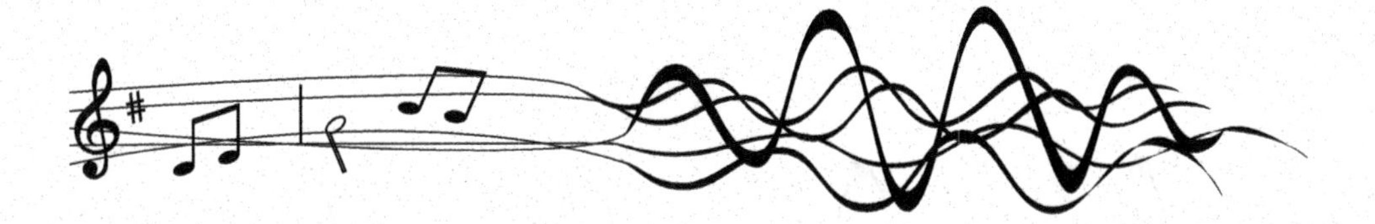

13.1 Air Column Vibrations

The strings discussed in Chapters 11 and 12 were the first example of a vibrating system that has a harmonic spectrum. Another system, which has been widely used by humanity to create musical instruments, is also unbelievably simple: tubes.

Let us consider so far only tubes with uniform round cross-sections, which are called cylindrical. The real instruments rarely have exactly cylindrical pipes, for instance, organs can even have square pipes made of wood. Also some pipes have gradually changing cross-sectional areas, which make them conical. But for now this is a sophistication we do not need to discuss.

13.1.1 The Tube with Both Ends Open

For starters, let us discuss the spectrum of a simple tube with both ends open, the sort of chimney pipe. When warm air is blown into this pipe, it may produce a low-frequency humming sound. If you take a piece of any metallic tube and tap on it, you will also hear a sound of definite frequency. This sound is not what we call musically valuable but at least the pitch is detectable and pretty steady.

Air can easily go in and out from the end of such pipe, and this immediately shows us the main difference between such a pipe and the string considered in a previous chapter. The string has displacement nodes at the end because the ends are fixed, meanwhile, the tube should have displacement antinodes at the end. Fig. 13.1 a shows the sketch of the movement of the air inside such tube. The closer we are to the end of the pipe, the bigger the amplitude of the vibrations of

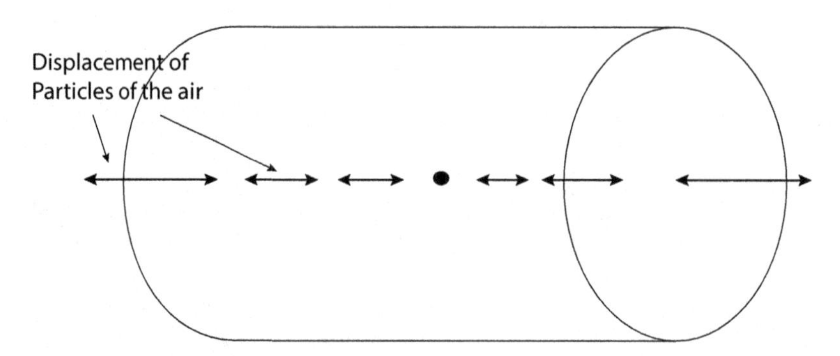

Fig. 13.1a. Tube open from both ends. Displacement of particles of the air is shown.

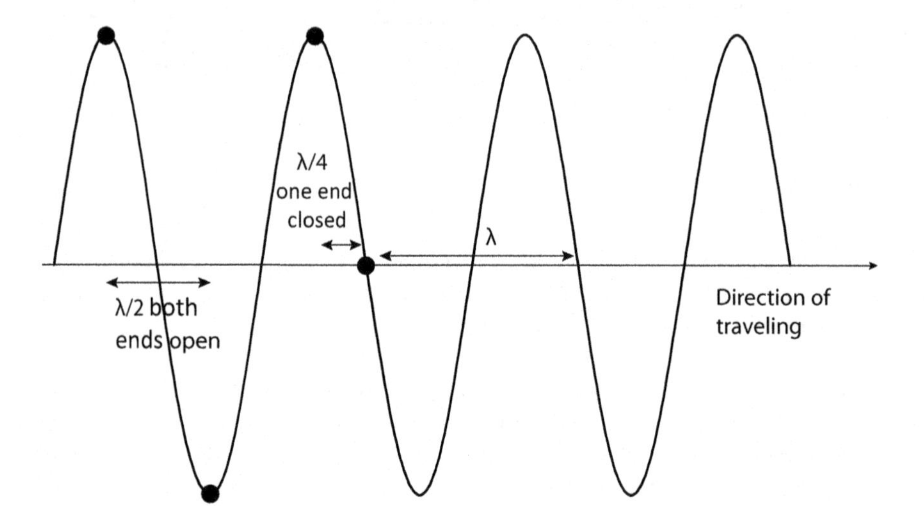

Fig. 13.1b. Traveling wave: the lengths of wavelength, half and quarter of wavelength are shown for illustration.

the air molecules is. Let us discuss this figure. In the middle, air is practically at rest (the nodal point), and at the ends the air is moving back and forth with greatest amplitude (antinodes). These points are stable, which means that we again created a standing wave.

To understand what fraction of the traveling wave we have now, check Fig. 13.1b, which shows the traveling wave. Now we have a part of the wavelength from maximum to negative maximum of the displacement, which corresponds to half of the wavelength of the traveling wave.

$$L = \frac{\lambda_1}{2}$$

where L is the length of our pipe. That's it: the length of the pipe with both ends open corresponds to half of the wavelength of the traveling wave. The frequency of the first normal mode can be written as

$$f_1 = \frac{V_s}{\lambda_1} = \frac{V_s}{2L}.$$

where v_s is the speed of sound in air. Note: this is important: for a string we had the speed of the wave in a string, which was defined by elastic and inertia properties of a string; now we have just the speed of sound in air because it is created in air.

The second harmonic of the tube with both ends open can be formalized the same way as we have done for a string:

$$L = \lambda_2$$

$$f_2 = \frac{v_s}{\lambda_2} = \frac{v_s}{L} = 2f_1$$

Now we have three antinodes (two at the edges, one in the center) and two nodes at distance L/4 from the edges. The first three normal modes for a tube open from both ends are shown in Fig. 13.2. The general formula for the frequency of the mode number n looks like:

$$\lambda_n = \frac{2L}{n}$$

$$f_n = \frac{v_s}{\lambda_n} = n\frac{v_s}{2L} = nf_1$$

The only difference between the string and the tube is that for the tube we always have antinodes on the ends, so the nth mode will have n+1 antinodes and n nodes.

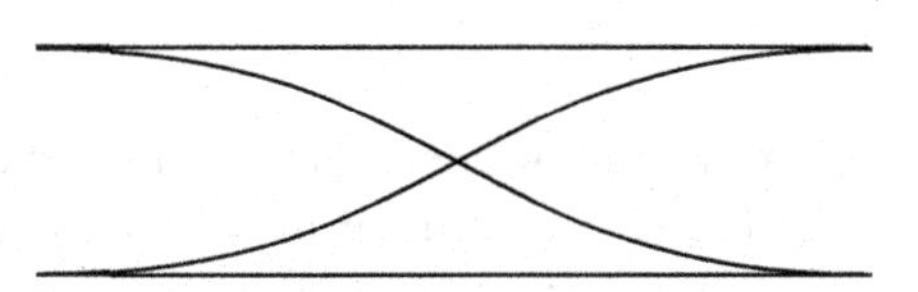

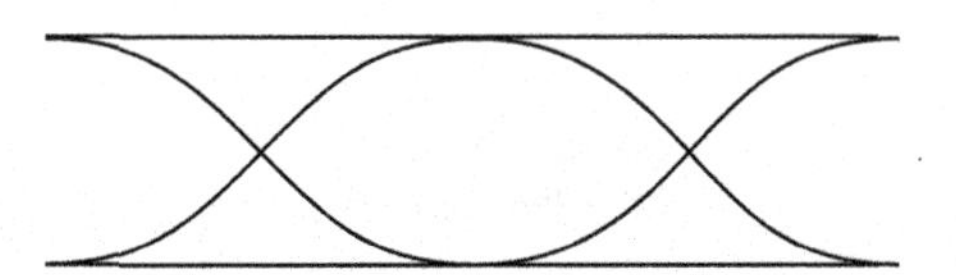

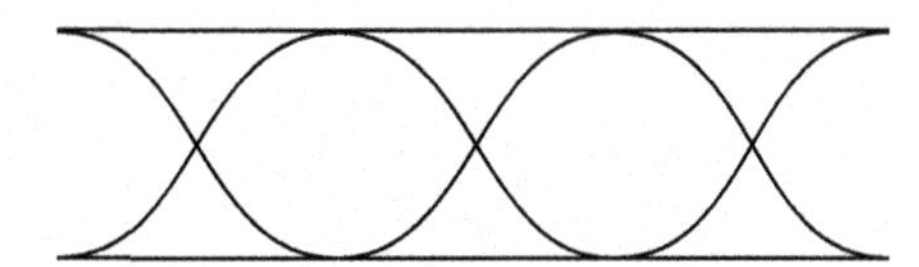

Fig. 13.2. First three normal modes of a tube open from both ends.

Example 13.1. What is the frequency of the fundamental mode for a tube of length 1 m at temperature 20°C? What will the frequency of this mode be at 30°C?

Answer:

The speed of sound at 20°C is 344 m/s. The fundamental of the tube of length 1 m is: 344 m/s/(2 x 1m) = 172 Hz. The speed of sound at 30°C is 350 m/s, and the corresponding fundamental will be 350 m/s / (2 x 1 m) = 175 Hz.

Example 13.2. What are the wavelengths of the first three modes of the tube of length 120 cm?

Answer:

The wavelength of the fundamental for this pipe is 2 x L = 120 cm x 2 = 240 cm. The length of next mode is *2L/2 = 120 cm*, the wavelength of third one, *2L/3 = 80 cm.*

To conclude, the tube open from both ends demonstrates a naturally harmonic spectrum and thus is another promising element for the creation of musical instruments. The tube with both ends closed, in principle, will have the spectrum analogous to the one considered above but with two nodes at the ends. However, such an element holds no musical interest.

13.1.2 The Tube with One End Closed

Quite often musicians use the term "closed pipe," and it can be confusing because what they mean is a tube open on one end and closed at the other. The model of such a tube you may create by yourself with just a plastic bottle with a wide opening. When you blow air across the opening, you hear a sound of a detectable pitch. The bottom of the bottle works as the closed end of the tube.

How air vibrates inside of such a tube is shown schematically in Fig. 13.3. At the open end we again have an antinode: air moves back and forth with a big amplitude. At the closed end air cannot oscillate: there is a wall, a node. Comparison with Fig. 13.1b shows that this fraction of a traveling wave is now one quarter—the distance from maximum displacement to zero.

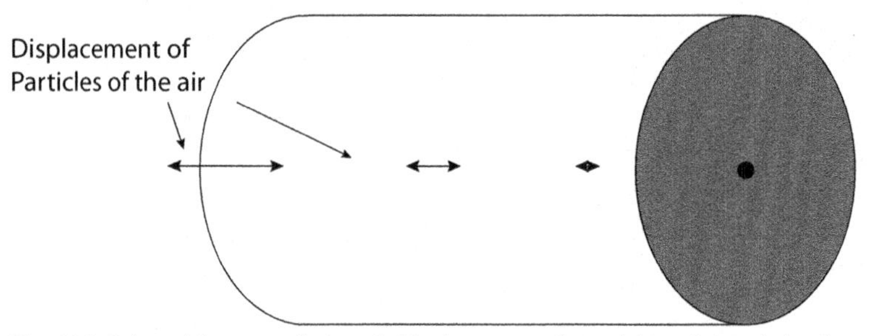

Fig. 13.3. Tube with one end closed. Displacement of particles of the air is shown.

$$L = \frac{\lambda_1}{4},$$

$$f_1 = \frac{V_s}{\lambda_1} = \frac{V_s}{4L}$$

We can immediately make a conclusion about what would happen to the voice of a tube if we close one end of a tube with two ends initially open. As it can be seen from a comparison of formulas for the fundamental mode of these tubes, the closing of one end will lead to the lowering of the voice of the tube by an octave.

Three first normal modes of the tube closed at one end are shown in Fig. 13.4. Note that because of the necessity to have a node at one end and an antinode on the other, these modes fit 1/4, 3/4, 5/4 of the wavelength. We always have an extra quarter of the wavelength. We may say that the spectrum of the tube closed from one end has only odd harmonics:

$$\lambda_n = \frac{4L}{n};$$

$$f_n = \frac{V_s}{\lambda_n} = n\frac{V_s}{4L} = nf_1$$

where n now is **only** odd.

Example 13.3. The length of some pipe is 120 cm. What are the frequencies of the first three modes at temperature 20°C if the tube has both ends open? If the tube has one end closed?

Answer:

At 20°C the speed of sound is 344 m/s. The frequency of the first mode for the pipe with both ends open is 344/(2xL) = 344/ (2 x 1.2 m) = 143.3 Hz. The frequency of the second mode is twice as big, 285.6 Hz, as the third mode—three times as big, 430 Hz.

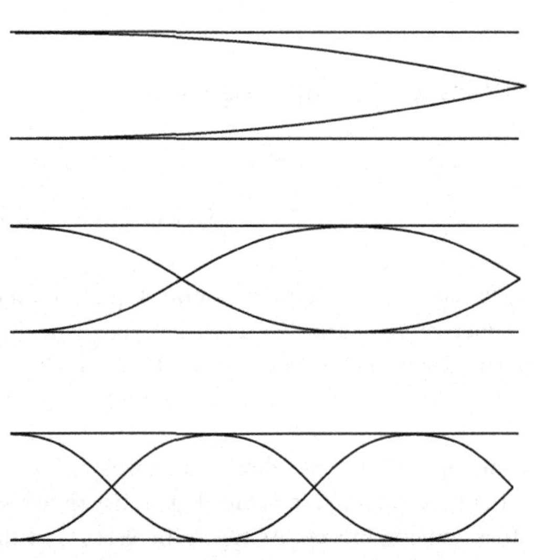

Fig. 13.4. First three normal modes of the pipe with one end closed.

If one end of the tube is closed, the frequency of the first mode is 344/(4 x L) = 344 m/s / (4 x 1.2) = 71.6 Hz. The next mode will be the third harmonic because even harmonics are absent in the spectrum of the pipe with one end closed. The frequency of the third mode is three times the frequency of the fundamental: 71.6 Hz x 3 = 214.8 Hz. The next mode will be the fifth harmonic with frequency 71.6 x 5 = 358.0 Hz.

Example 13.4. Some tube resonates at frequencies 200, 300, 400 Hz and nothing in between. What is the fundamental frequency of this tube? Does this tube have both ends open or one end closed?

Answer:

The fundamental frequency of this tube is the biggest common divisor of numbers 200, 300, 400: 100 Hz. As we see, in the spectrum of this particular pipe there exist both odd and even harmonics; hence this pipe is open from both ends.

Example 13.5. Some tube resonates at frequencies 300, 500, 700 Hz and nothing in between. What is the fundamental frequency of this tube? Does this tube have both ends open or one end closed?

Answer:

The fundamental frequency of this tube is the biggest common divisor of numbers 300, 500, 700: 100 Hz. In the spectrum of this pipe only odd harmonics are present; hence this is a pipe with one end closed.

Now we can understand better why the Fletcher-Munson diagrams demonstrate high sensitivity of the human ear at 3 to 4 kHz. This is merely a resonance effect in the ear canal of the outer ear. The canal is roughly 2.5 to 3 cm long, and it is just a tube with one end open, another closed. The first normal mode of such a tube should have the wavelength 10–12 cm, which corresponds to a frequency of 3.5 kHz at reasonable temperatures. The next normal mode will be at 10.5 KHz, for which our ear is also sensitive, but this frequency is already more than one octave higher than the extreme treble of the piano.

13.2 Fluid Jets and Edgetones

13.2.1 Laminar and Turbulence Flow

In Physics, "Fluid" applies to both liquid and gas. Just as mixing coffee inside of a cup with a spoon creates little "vortexes," so does plucking or hitting a string cause air vibrations. For horned instruments, the tube itself has a harmonic spectrum, and can serve as a resonator for music. These vibrations caused are fluid-flow instabilities.

Laminar, or "layered," flows, as shown in Fig. 13.5a, are smooth, steady, and observed only if the speeds are quite small. Fig. 13.5a shows that the layers of fluid do not mix with each other, moving in step. As speed increases, and the fluid approaches **critical speed**, oscillatory disturbances grow larger. More energy is needed for these larger disturbances, and thus the higher the speed, the less steady and orderly the flow (Fig. 13.5b). A flow might still be fairly steady if the speed is just around or past critical speed, but if it exceeds it greatly, the number of oscillatory disturbances of many wavelengths and frequencies grows so much that it can no longer be stable. This is called **turbulent** flow.

Unstable fluid flows serve as the source of sound.

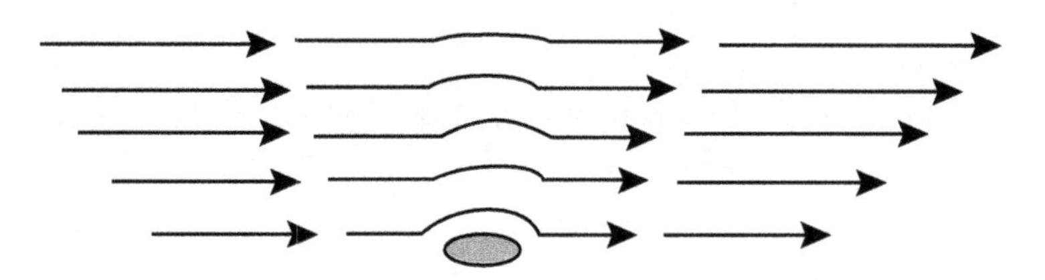

Fig. 13.5a. Example of laminar flow of a fluid: the "layered" character of the motion can be seen.

13.2.2 Fluid Jets

Blowing air through a narrow tube at a slow enough speed (laminar flow) will produce very little noise (Fig. 13.6a). As the speed increases (turbulent flow), the sound increases (Fig. 13.6b). The higher the speed, the less stable the slow, the greater the sound. If no periodic disturbances, or strict periodic oscillations, are present, however, there will be no definite pitch.

Fig. 13.5b. Example of turbulent flow of a fluid: all layers are mixed, some vortexes can be seen.

13.2.3 Edgetone

If we introduce a sharp-edged obstacle directly in the way of air coming from a narrow slit (this narrow slit is called a **flue**), (Fig. 13.7), there will be many more vibrations concentrated mainly about one frequency, resulting in louder sound with a more definite sense of pitch. This is creating an **edgetone.**

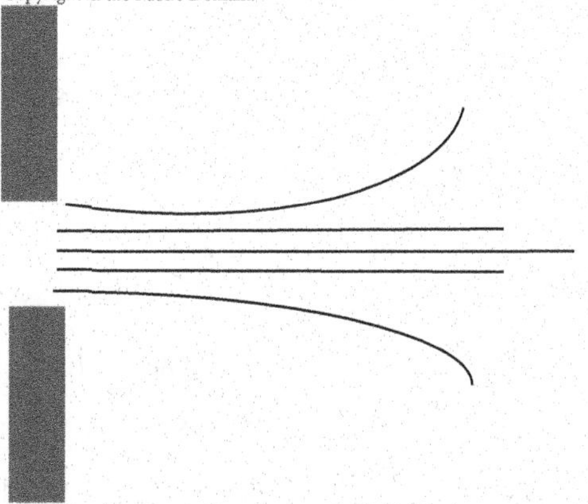

Fig. 13.6a. Air coming from the narrow opening at the small speed of the jet: almost laminar motion.

Whistle by using a piece of paper; your mouth serves as a narrow opening, and the air from your mouth is broken against the piece of paper. The frequency of this edgetone changes based on positioning of the paper; increasing distance between your mouth and the edge lowers the pitch, increasing the strength of blowing—increasing the speed of the air—rises pitch. This demonstrates that the pitch of the edgetone is dependant on speed, as shown in Fig. 13.8a.

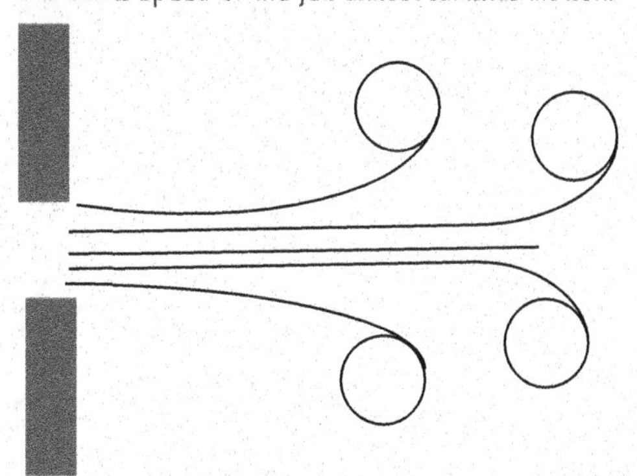

Fig. 13.6b. Air coming from the narrow opening at the speed of the jet bigger than critical: several vortexes are shown.

In this situation, the distance from the opening to the edge, called a gap, is fixed. The speed of air increases the pitch, but in a pretty sophisticated way: there are some speeds for which edgetones can operate in either of two stages with different frequencies. The sudden jumps between stages are indicated with arrows.

Fig. 13.8b shows the dependence of the edgetone pitch on another parameter—gap size with a constant air speed. The increase of the gap distance b lowers the pitch, but again at a certain point the frequency demonstrates sudden jumps. For normal musical applications, only the first of all these stages is important; others are extremely unreliable.

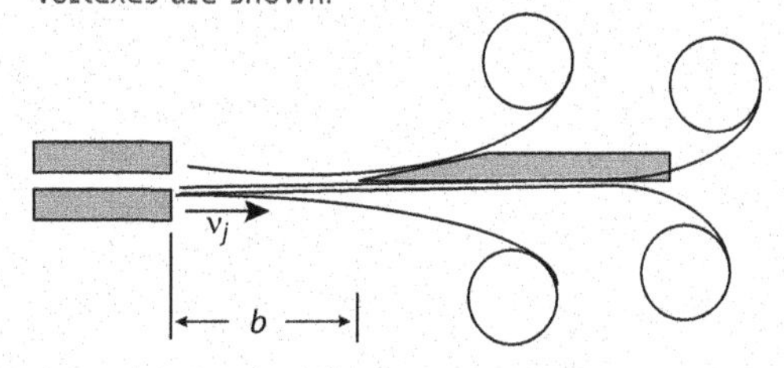

Fig. 13.7. Edge is positioned in the way of the airflow from the narrow opening.
Source: Musical Acoustics by Donald Hall, 2002 Cengage Learning, Inc.

Fig 13.9 shows the air forming vortexes on one side and another before reaching the edge. The main flow is slightly wavy with almost circular turbulences. The edge divides one region of space from another, actively participating in the process. When the air begins to flow below the edge, it causes higher fluid pressure, pushing up on the edge. This higher pressure forces the other fluid to move out; part of this pushed out flow goes up through the gap between the air and flue (Fig. 13.10). It also pushes upward on the air, so that a short time later the air will flow above the edge instead of below.

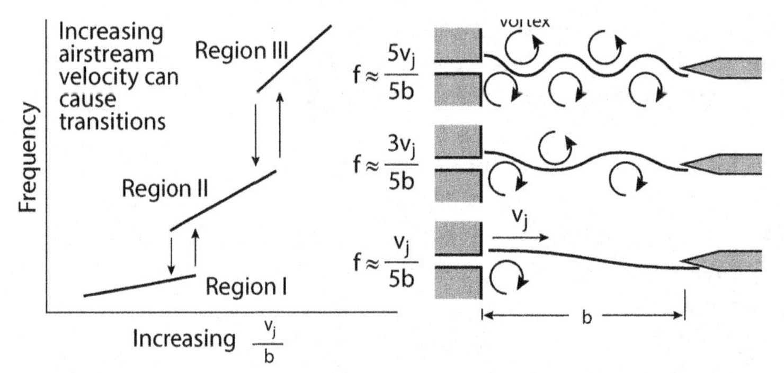

Fig. 13.8a. Dependence on the frequency of an edgetone on the jet speed (left). On the right, the corresponding profile of the airstream for each region is shown.
Source: Musical Acoustics by Donald Hall, 2002 Cengage Learning, Inc.

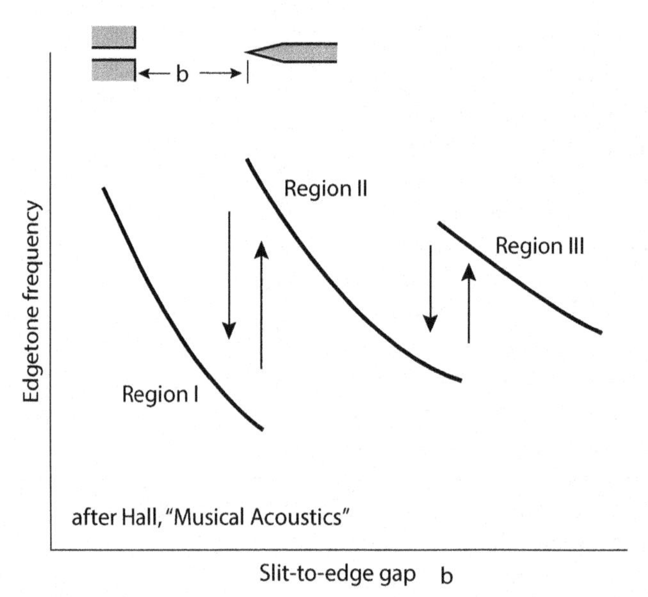

Fig. 13.8b. Dependence on the edgetone frequency of the gap between the narrow opening and the edge.
Source: Musical Acoustics by Donald Hall, 2002 Cengage Learning, Inc.

This is called positive feedback: the pressure created by the interaction of the air with the edge feeds back to the area of the slit, tending to push the stream upward. The reverse happens when the stream moves to the top side of the edge and then the process repeats itself.

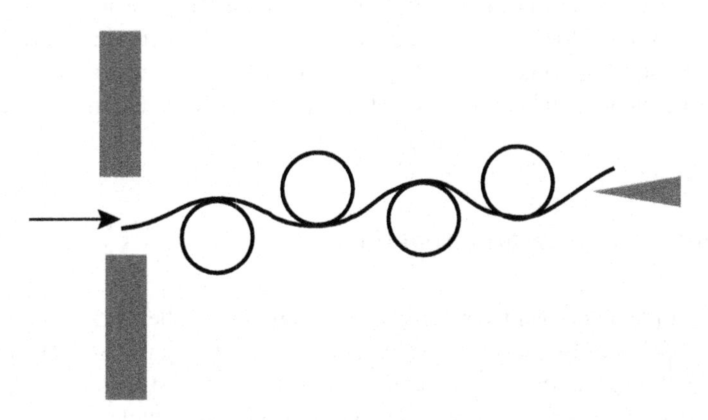

Fig. 13.9. Jet forming vortexes before it reaches the edge.

As a result, a periodic flipping of the airstream from side to side is created.

The frequency of oscillation of the edgetone is controlled by the speed with which the vortex travels across the gap. This speed is approximately 0.4 times the speed of jet itself. This disturbance actually moves upstream of the jet while the jet carries it to the edge. So, half of the period of oscillations should be the time it takes to cover the gap distance b at speed of the vortex 0.4 v_j:

$$\frac{T}{2} = \frac{b}{v_j}$$

Then the preferred frequency of stage one is

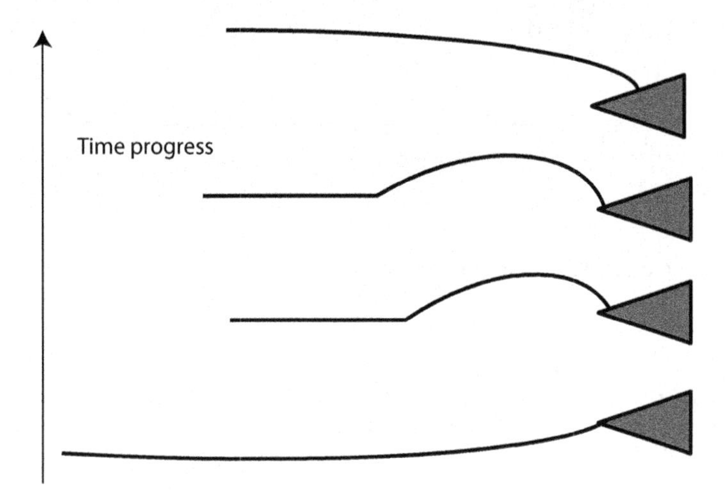

Fig. 13.10. Airflow pushes jet up or down as time progresses.

$$f_1 = \frac{0.2v_j}{b}$$

This formula gives the dependence of the first mode of the edgetone, shown in Fig. 13.8a, b on the speed of the vortex and gap distance. The switch to higher frequencies can be explained by considering the possibility that the vortex can take 3/2 or 5/2 of a period to reach a gap instead of 1/2. Each of these will also generate a positive feedback and maintain oscillations at $f_{II} \approx 3f_1$ and $f_{III} \approx 5f_1$. The frequencies in between will not create positive feedback and will be suppressed.

13.3 Edgetone and Resonator Together

At first, it is not a difficult problem to connect an edgetone and a tube. We know the spectrum of a tube, we know the lowest frequency of edgetone, which means that we should take an edgetone with the frequency closest to the fundamental of the tube and that's it (Fig. 13.11). Alas, the tube will not just support the frequency of our edgetone, it will actively influence it.

The resonance inside of a tube builds up much bigger pressures and stronger acoustic flows than the isolated edgetone would, and thus the tube performs strong control over the edgetone. Specifically, the presence of the resonator makes the whole system speak at a lower frequency than would the isolated edgetone. The tube provides its own strong positive feedback in the form of a wave that travels to the far end of the pipe, reflects back, and disturbs the jet upon its return. The more detailed consideration, which is beyond the scope of this book, is that **the attached resonator lowers the normal frequency of the edgetone by an octave**. The frequency of the

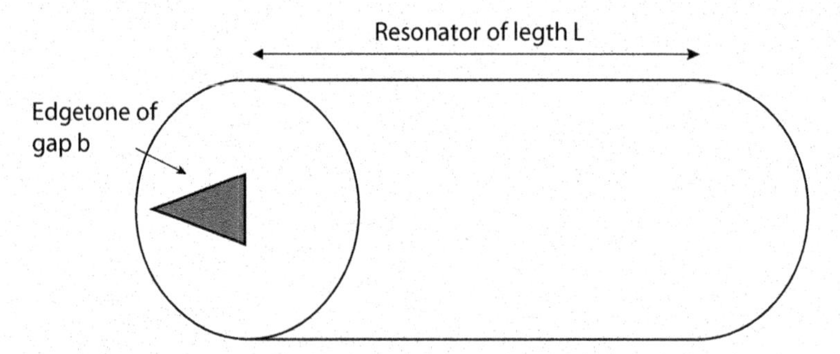

Fig. 13.11. Sketch of edgetone with gap b and resonator tube of length L.

fundamental mode of the edgetone with a resonator is

$$f_1 = \frac{0.1v_j}{b}.$$

Example 13.6. The gap distance for some edgetone is 0.5 cm, and the speed of jet is measured to be 100 m/s. What is the frequency of the fundamental mode of this edgetone?

Answer:

The fundamental of the edgetone is defined from the formula:

$$f_1 = \frac{0.2v_j}{b} = \frac{0.2 \cdot 100m/s}{0.005m} = 4000Hz$$

Example 13.7. What frequency will produce the edgetone from the previous question when connected to a resonator?

Answer:

The resonator lowers the frequency of the edgetone by an octave, hence the edgetone from the previous example will have fundamental frequency 2000 Hz.

Example 13.8. If we connect the edgetone from the previous question to a pipe open from both ends, what is the length of this pipe? If this pipe has one end closed, what would be its length? Assume temperature 20°C.

Answer:

This pipe should have the fundamental frequency 2000 Hz; the speed of sound at the given temperature is 344 m/s. Hence, the length of this pipe can be found from the formula 2000Hz = 344 m/s / (2 x L), so L = 344 m/s / (2x 2000 Hz) = 0.086 m = 8.6 cm.

13.4 Organ Flue Pipes

The mouth of a typical organ flue pipe is shown in Fig. 13.12. In normal operation the pipe Mostly produces the lowest normal mode. Sound outside of the pipe is different in color in comparison to what is inside because the radiation of upper harmonics is more efficient. The typical spectra of several flue pipes are shown in Fig. 13.13. For some forms of pipes, upper harmonics may successfully rival fundamental in amplitude.

The differences in these spectra can be explained, first of all, by differences in pipe proportions. Flue pipes can be classified as: flute, which are rather fat; string, which are skinny; and principal, whose thickness is in between.

The fat, wide pipes usually are poor in upper harmonics. The formula that we discussed in the beginning of the chapter for a cylindrical

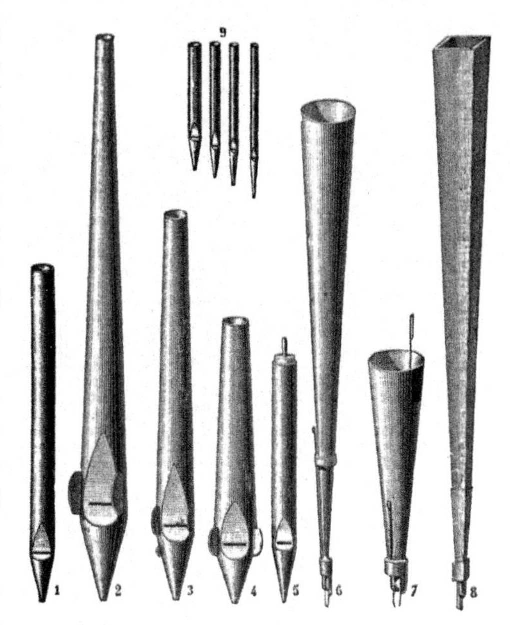

Fig. 13.12. Organ flue pipes of different configurations.

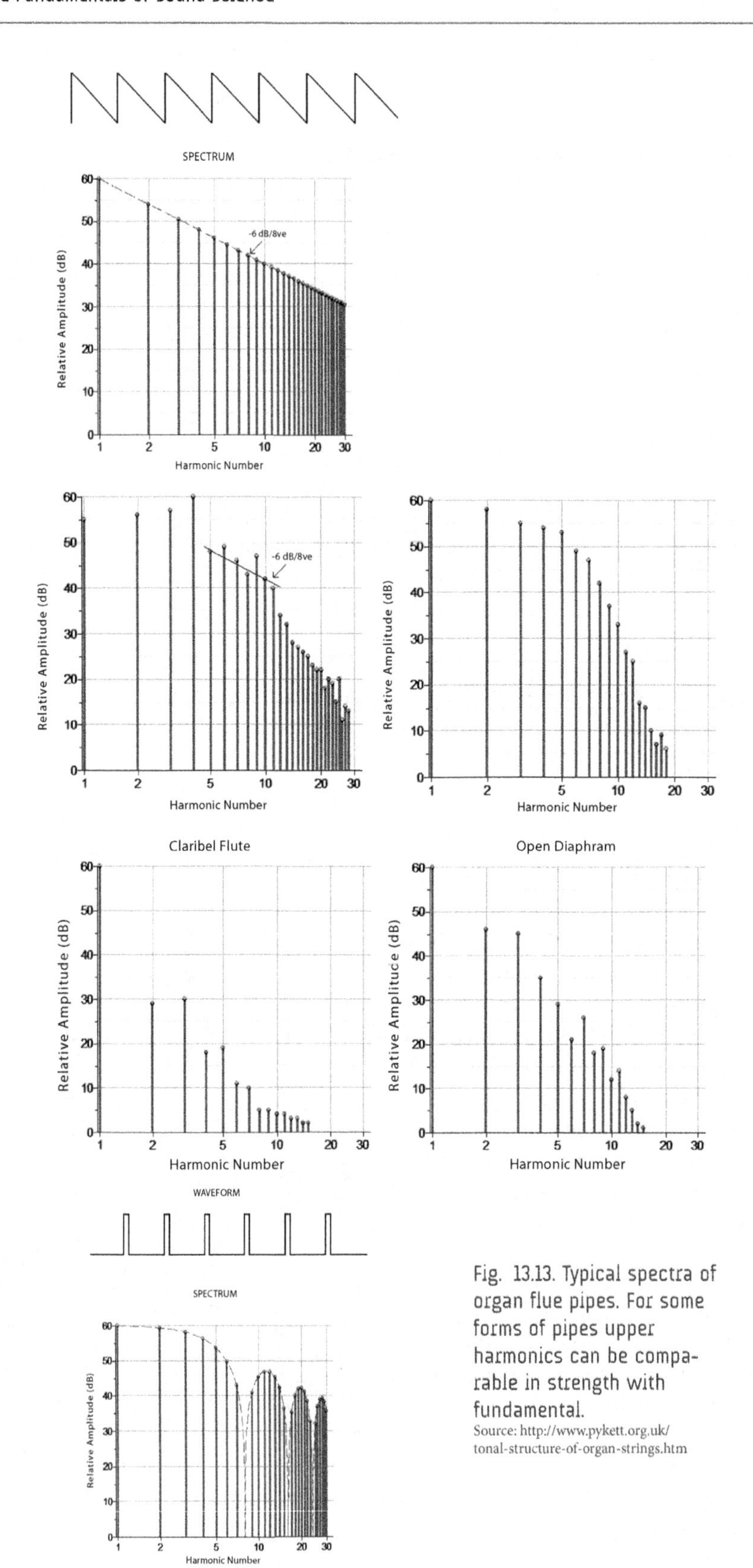

Fig. 13.13. Typical spectra of organ flue pipes. For some forms of pipes upper harmonics can be comparable in strength with fundamental.

Source: http://www.pykett.org.uk/tonal-structure-of-organ-strings.htm

pipe is valid only for very thin, skinny pipes. The wide pipe can keep inside only the sound of pretty big wavelengths. Actually, the sound of a wavelength bigger than half the diameter of a pipe is reflected back from the end pretty effectively. But for small wavelengths the open end of wide pipe is just invisible—they escape from a pipe with no strong standing wave established. Behind some critical frequency, higher harmonics will be absent from the musical recipe. The good estimation of the number of effective harmonics is

$$N = \frac{L}{D}$$

where L is the length of a pipe and D its diameter.

A complete set of pipes producing similar tone color is called a **rank**. Most ranks have 61 pipes (five octaves), and the longest should be 32 times longer than the shortest one. To make a rank the organ builder should solve a problem of scaling, deciding how the diameters and lengths should vary for different pitches. This problem is not as easy at it seems because simply keeping the diameters the same or keeping the ratio L/D the same does not provide a uniform and rich voice for all frequency ranges. The outstanding organs usually have the bass end of spectra richer in harmonics than the treble. The diameters for these pipes change a little more slowly than the lengths. Trial and error shows that the best ratio of diameters between the largest and smallest pipes is 12 instead of 32.

13.5 Fingerholes

Organs are instruments in which one pipe is responsible for only one note. There are also other, more portable, instruments which also use one pipe per note; for instance, the panflute (Fig. 13.14). A continuous change of notes can also be obtained by changing an air column length, for instance, as in slide whistles.

But there is also a possibility to change the tone without switching to another pipe or even without mechanical change of its length. What effect should a single large hole in the side of a pipe have? Let us consider Fig. 13.15a. With the hole closed, the fundamental mode of the pipe uses the opposite open end to vibrate. But with the hole open there is no longer a solid wall, so little airflow will run to the end of the pipe with the possibility of going in and out of the hole. So, the open hole serves now as the displacement antinode, effectively decreasing the length of the pipe and raising pitch. The closer the hole is to the blowing end, the shorter the wavelength of the sound will be.

This picture is oversimplified for real life. It is pretty rare when the holes are big enough to allow air to move back and forth as it does from the end of the pipe. The smaller the fingerhole, the less airflow escapes from it. A small hole closer to a blowing end can produce the same effect as a large hole farther down. (Fig. 13.15b, c). To play an ascending scale we can start with all holes covered and lift one finger at a time starting from the far end of the pipe.

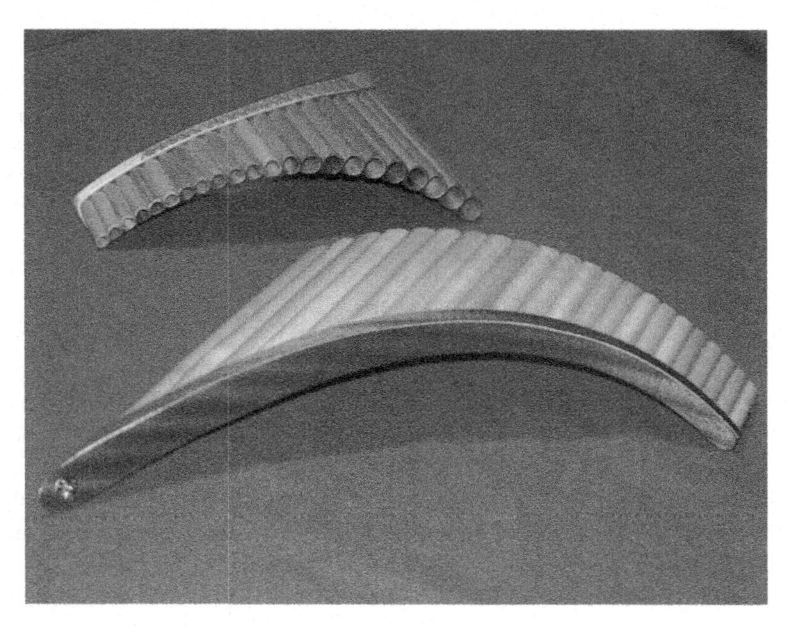

Fig. 13.14. Panflutes.

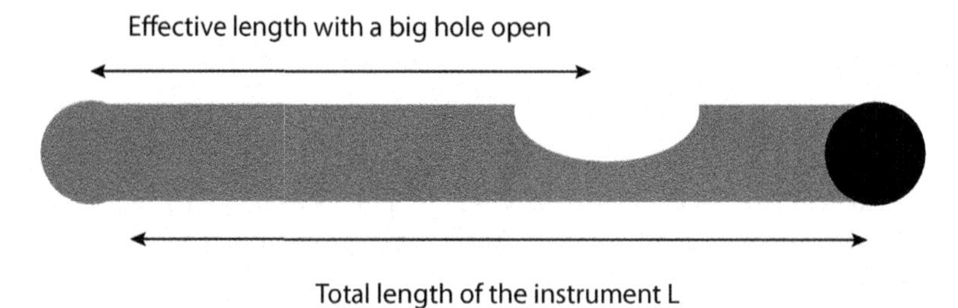

Fig. 13.15a. Influence of the hole on the pitch of the instrument: with hole open, the effective length of the instrument becomes shorter.

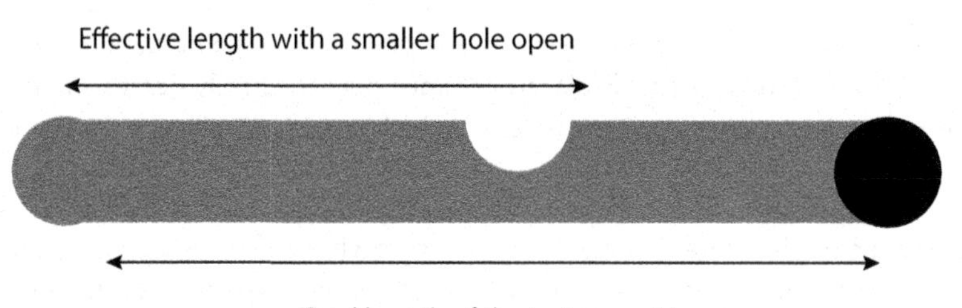

Fig. 13.15b. Effective length of the instrument with a smaller hole than in Fig. 13.15a hole open is longer than in Fig. 13.15a.

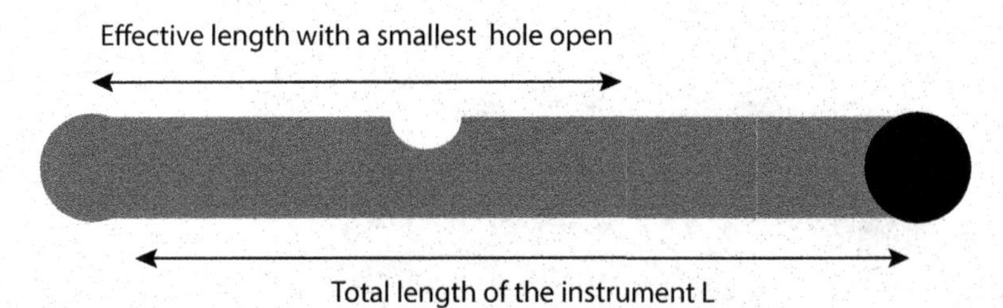

Fig. 13.15c. Opening of an even smaller hole than in Fig. 13.15b: effective length of the instrument is shortened less than in Fig. 13.15b.

13.6 Recorders and Flutes

The recorder is a very old instrument, initially made from a straight branch of a tree (Fig. 13.16). It has cylindrical body with 7 holes with diameters 30–50% of the diameter of pipe. The recorder's fipple mouthpiece (Fig. 13.17) has a fixed gap, and a player can control only the speed of the jet. Only skilled players can complete even the second octave on a recorder, so the range of the instrument is pretty narrow. The recorder was revived in the twentieth century, partly for the performance of early baroque music, because of its suitability as a simple instrument for teaching music, and for its appeal to amateur players. Today, it is often thought of as a child's instrument, but there are many professional players who demonstrate the instrument's full solo range. The sound of the recorder is very soft, clear, and sweet.

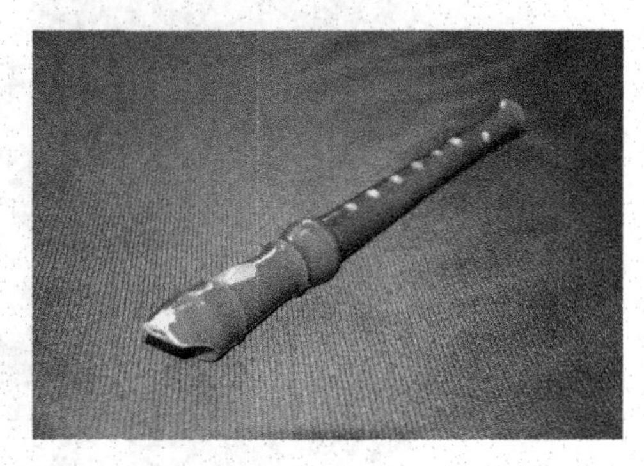

Fig. 13.16. Recorder.

The transverse flute is another edgetone instrument, used today much more often than a recorder (Fig. 13.18). It has 12 holes, which correspond to the 12 notes of the first octave. The holes are larger than a recorder's: 70–80% of the diameter of a pipe. Some holes are "normally closed" and some "normally open," which allows a performer to operate 12 holes with only 8 available fingers. The flute has an adjustable gap, which is created by the configuration of a performer's lips (Fig. 13.19). This gives a flutist an advantage over the recorder player in being able to change the gap distance by moving the lips or rolling the instrument up or down. Having control of the jet speed and gap distance, the flutist is able to change in various ways timbre, loudness, and pitch. As a result, even a third octave is accessible on the flute.

Fig. 13.17. The edgetone of a recorder: fixed gap.

Fig. 13.18. Flutes.

Fig. 13.19. The flute has an adjustable gap: the size of the gap can be controlled by the configuration of the performer, more portable, 's lips or by rolling the flute slightly up or down.

Meanwhile, the flute is an extremely "inefficient" instrument: only 0.1% of energy spent by the player's lungs converts into sound. In the last 30 years the flute has often been used in hard-rock compositions. For instance, the voice of flute is a hallmark of the British group "Jethro Tull," with a lead flutist of unbelievable skills, Ian Anderson. Because of the general loudness of rock performances, he uses his flute only with a powerful amplifier.

Summary, Terms, and Relations

The spectrum of the pipes with both ends open:

$$\lambda_n = \frac{2L}{n} \qquad f_n = \frac{v_s}{\lambda_n} = n\frac{v_s}{2L} = nf_1$$

The spectrum of the pipes with one end closed:

$$\lambda_n = \frac{4L}{n}$$

$$f_n = \frac{v_s}{\lambda_n} = n\frac{v_s}{4L} = nf_1, \text{ where n is only odd.}$$

The fundamental frequency of an edgetone:

$$f_1 = \frac{0.2v_j}{b}$$

The fundamental frequency of the edgetone connected to a resonator:

$$f_1 = \frac{0.1v_j}{b}$$

Questions and Exercises

1. For an open tube of length 60 cm, what are the wavelength and frequency of the fundamental mode?

2. For a closed tube of length 1.6 m, what are the wavelength and frequency of the fundamental mode? Of the third mode?

3. What would happen to the voice of an open-ended tube if you close one end?

4. For a fundamental frequency of 740 Hz, what length would be required on an open tube? What length on a closed tube?

6. Some organ pipes resonate at frequencies 400, 500, and 600 Hz and nothing in between. Is this tube open or closed at one end? What is the fundamental frequency of this tube?

5. If a simple edgetone setup has gap 0.4 cm and jet velocity 35 m/s, what would be the approximate frequency of the edgetone?

7. If the edgetone from the previous question is connected to an open end tube, what length of tube is required?

8. Two open pipes both have length 50 cm, but pipe A has diameter 10 cm, and pipe B diameter 2 cm. What is the number of harmonics that you could expect in the spectrum of each?

9. How does a large hole that is drilled on the side of a tube influence the voice of this tube?

10. Suppose you have a 66 cm open pipe that produces a C_4. You drill a large hole 44 cm down from the blowing end. What note do you expect with the hole open?

Chapter 14

The Reed Family

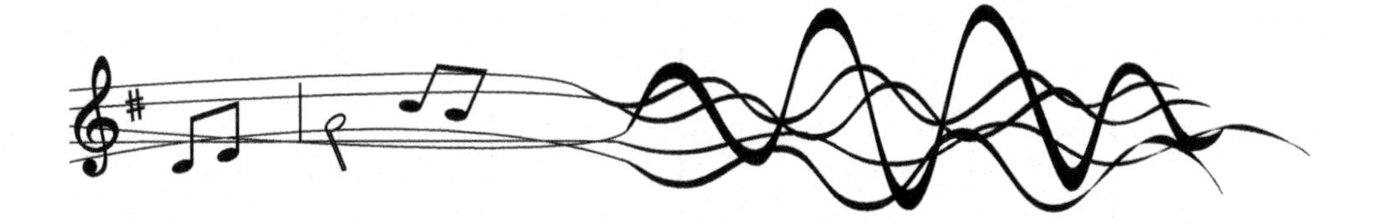

14.1 The Reed

Many instruments with a tube as the resonator use a small device, called the reed, for the creation of steady standing waves. Some organ pipes in a good organ, clarinets, saxophones, oboes, and trumpets use reeds of different shapes (Fig. 14.1).

First we should discuss the Physics behind the motion of a reed. What effect leads to the appearance of steady oscillations of the petal(s) of the reed?

For starters, we perform a simple experiment. Let's take a couple of sheets of paper, as shown in Fig. 14.2a, and blow air between them. What are our predictions? How will these sheets move? My students who play woodwinds have no problems at all with answering this question, meanwhile, others have difficulties. These sheets of paper move toward each other and oscillate (Fig. 14.2b).

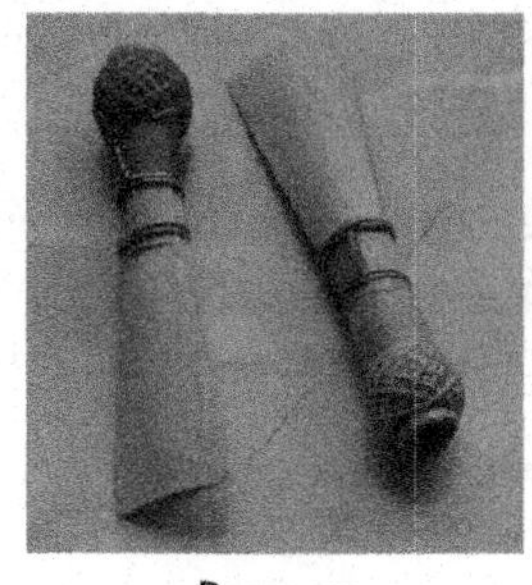

Fig. 14.1. Reeds of bassoon (left) and saxophone (right).

The effect that is responsible for oscillation of our "paper reed" is known as the Bernoulli effect, named in honor of the first scientist who described and explained it. So, what's going on?

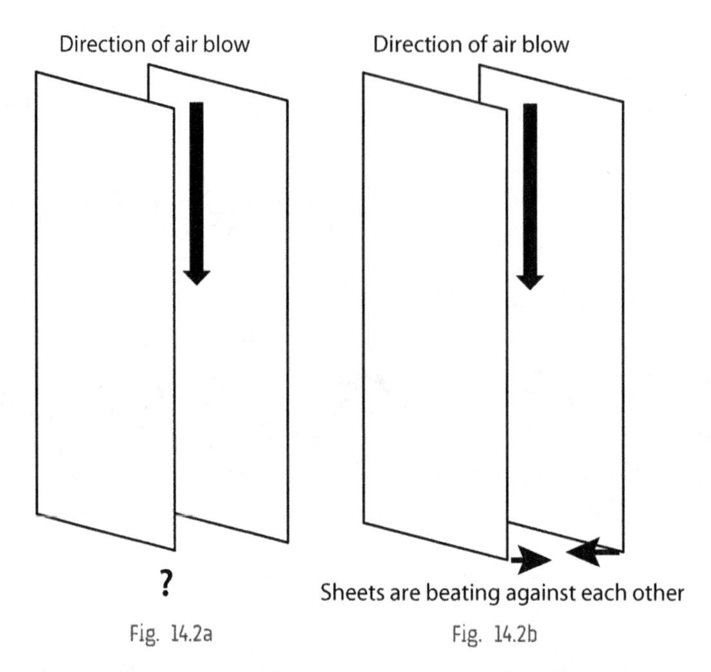

Fig. 14.2. Demonstration of Bernoulli effect: when air is blown between sheets, they start to beat against each other.

The Bernoulli effect, which is just a generalization of one of the basic principles of Physics, the law of conservation of energy, says: if in some part of a fluid the speed of this fluid is high, then the pressure in this fluid is low. And vice versa, if the speed is low, then the pressure is high.

Everybody has experienced this effect when we are trying to hide under an umbrella from the rain on a windy day, and the umbrella suddenly pops inside out or even breaks. The sole reason for it is the Bernoulli effect (Fig. 14.3). The speed of the air above the umbrella is bigger

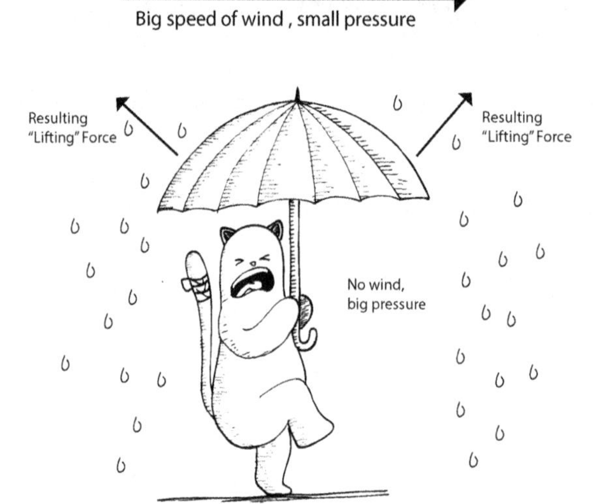

Fig. 14.3. Example of Bernoulli effect: the difference of pressure beneath and above the umbrella causes the umbrella to pop inside out.

than the speed of air under the umbrella. As a result, a net upward pressure is developed. The Bernoulli effect is the reason for the lifting force of aircraft and helicopters, as well as the lifting of roofs during hurricanes.

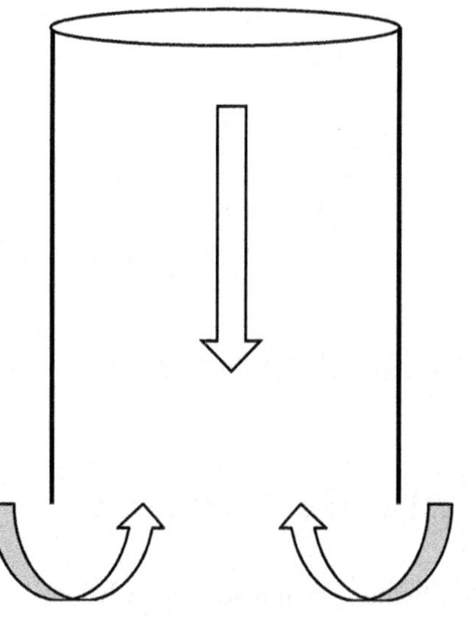

Fig. 14.4a. Motion of parts of double reed when the air is blown in between.

Exactly the same thing as we have seen for two pieces of paper happens with a reed (Fig. 14.4). For simplicity, a double reed is shown, but single reeds work the same way, beating against the edge of the mouthpiece of the instrument. When the speed of the blowing is small, the parts of the reed do not block the airflow, performing almost sinusoidal motion, like two pendulums (Fig. 14.4b). With increasing blowing speeds, the character of motion moves farther and farther from sinusoidal (Fig. 14.4c). As we know from Chapter 9, the

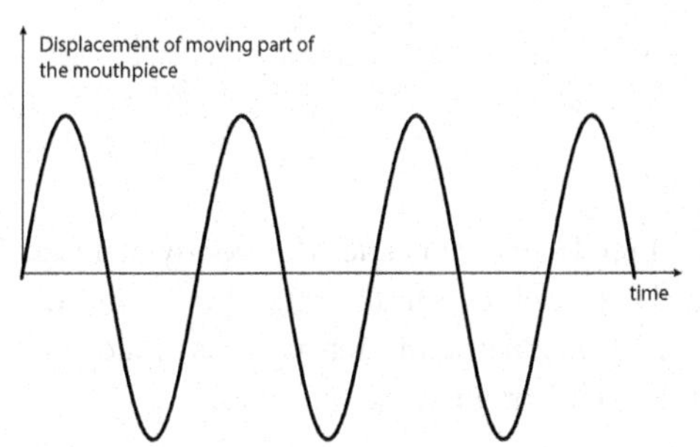

Fig. 14.4b. Moderate blowing speed: the motion of parts of the reed is almost sinusoidal with no blocking of the airway.

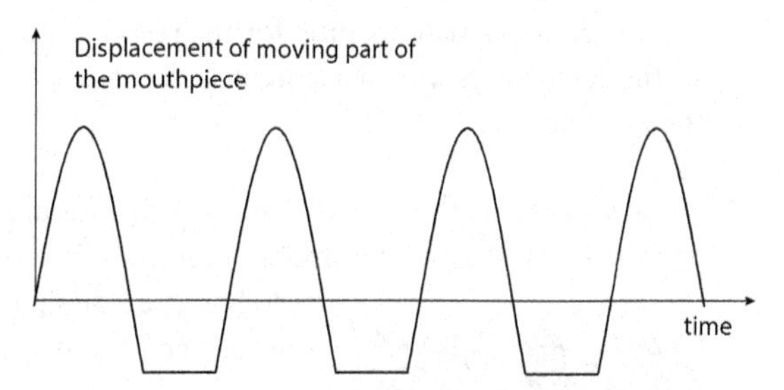

Fig. 14.4c. Big blowing speeds: the airway is blocked during the part of the period.

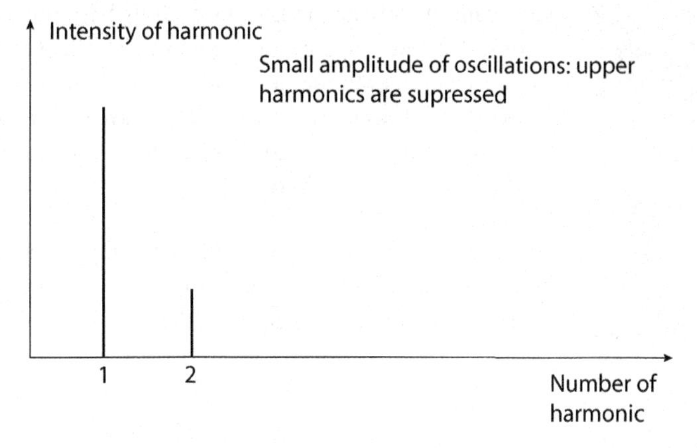

Fig. 14.4d. Spectrum of the sound produced by a reed at moderate blowing speed: upper harmonics are suppressed.

sinusoidal form has only one frequency of oscillations and no upper harmonics (Fig. 14.4d). Meanwhile, the form, shown in Fig. 14.4c, produces the whole set of overtones, which are delivered into the resonator of the instrument.

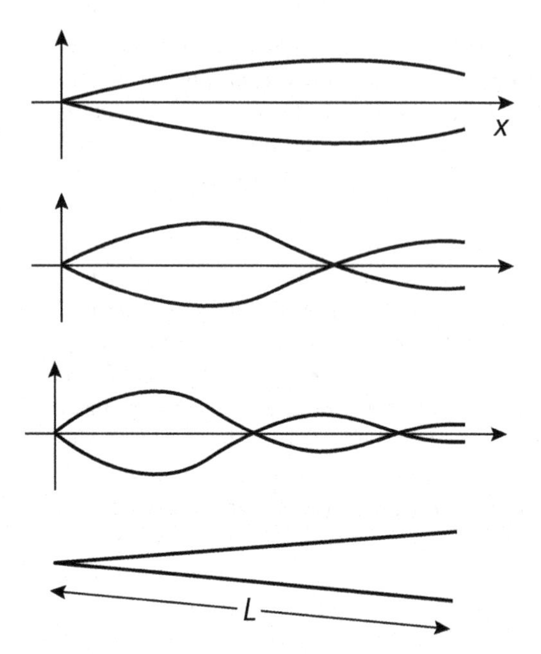

Fig. 14.5. First three normal modes of a conical tube.
Source: Musical Acoustics by Donald Hall, 2002 Cengage Learning, Inc.

14.2 The Reed Woodwinds

14.2.1 General Characteristics

All woodwind instruments have two common features: they all use a reed to create sound, and they all have a resonator tube, closed at one end. That's it: at the mouthpiece there is a small opening for the reed, so there is always a displacement node at the mouthpiece.

Usually the reeds that are used for woodwinds are soft reeds, made of cane. The woody material of cane is less elastic than metal, used mostly in reed organ pipes. On the other hand, the soft reed has more dissipation, the resonance of such a reed is broad, and the soft reed can respond equally on many frequencies.

Fig. 14.6. The clarinet.

But before we start the consideration of typical woodwinds, another problem that we know from Chapter 13 should be solved: the cylindrical tube, closed from one end, has only odd harmonics in its spectrum. As a result, many woodwinds have not a cylindrical, but a conical shape. The conical shape brings back even harmonics, and the wavelength of fundamental is twice the length of a pipe. Fig. 14.5 shows the displacement for a standing wave in a conical bore for first three harmonics. Such shape has harmonic spectrum with frequency $f_n = n(v_s/2L)$ of the nth harmonic.

Now we have two sets of possibilities to play with: single and double reed, cylindrical and conical bore. Let us create and discuss the properties of all possible combinations.

14.2.2 The Clarinet

The clarinet (Fig. 14.6) has a single reed and cylindrical bore with some flare at the lower end (Fig. 14.7a). Except for soft playing the reed beats

Fig. 14.7a. Small flare at the lower end of the clarinet.

Fig. 14.7b. The mouthpiece of the clarinet.

against the mouthpiece edge (Fig. 14.7b). The same name is used for the entire family of instruments of such features, with the B^b soprano being the most common.

The even harmonics are missing in the spectrum of the clarinet because of the shape of the bore. This explains why the voice of the clarinet stands apart more from other instruments. The timbre of a clarinet is a little bit hollow, especially in the clarinet's lowest notes.

The cylindrical, open-from-one-end tubes have another important feature: their length corresponds to one quarter of the wavelength of the sound. This explains why the clarinet always sounds an octave lower in comparison to a flute of the same size.

When a clarinet is overblown, the first available note is one octave and a perfect fifth higher (19 semitones). This means that a clarinet must have enough tone holes and fingering patterns to produce the lower register of 19 notes before overblowing begins.

14.2.3 The Saxophone

The saxophone (Fig. 14.8) has, like the clarinet, a single reed, but a conical bore. Here it is not so much of a bell shape as it is a small flange at the lower end of the instrument (Fig. 14.9). For saxophones the same name is also used for the whole family. For instance, the saxophone quartet is usually made up of one B♭ soprano, one E♭ alto, one B♭ tenor, and one E♭ baritone. Note that the angles of the cones of the saxophone bore are pretty small: 4° for soprano and 3° for tenor, which is more than enough for the appearance of even harmonics in the spectrum. As a result, the overblown notes of the saxophone sound one octave higher (12 semitones). The fingerholes can provide the lowest octave; all higher notes must be overblown.

Fig. 14.8. Differrent saxophones.

14.2.4 The Oboe and the Bassoon

The oboe (Fig. 14.10) and bassoon (Fig. 14.11) have double reeds and conical bores and so work

Fig. 14.9. Saxophones do not have a bell but rather a flare at the lower end of the instrument.

Fig. 14.10. Different oboes.

Fig. 14.11. Bassoon.

on the same acoustic principles. They are the soprano, alto, and bass of what we shall call the oboe family.

The oboe also has even and odd harmonics in its spectrum, but the angle of the conical bore is smaller than for the saxophone: 1.4° for the oboe and 0.8° for the bassoon. The special timbre of the oboe, in comparison to the wider saxophone, is caused by its narrowness that makes the voice rich in the highest harmonics. As we remember from Chapter 13, the highest harmonics easily escape from the wide open end of the instrument.

The overblown notes of the oboe are one octave higher than the basic register, like the case of the saxophone because of the existence of both even and odd harmonics.

Fig. 14.12. Crumhorn.

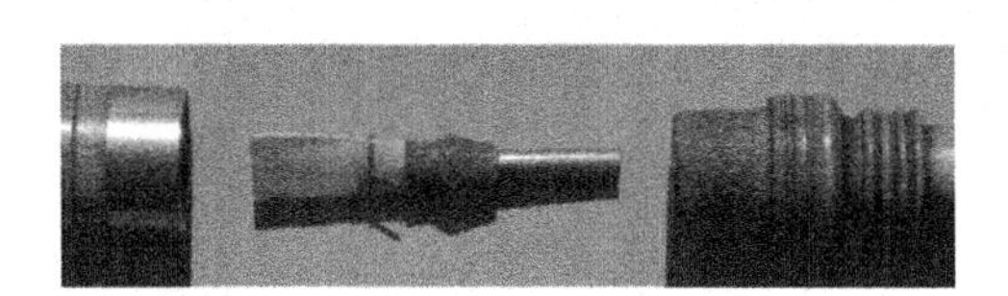

Fig. 14.13. The crumhorn's capped reed.

14.2.5 The Crumhorn

The crumhorn uses double reed and conical bore, and this is the fourth possible combination to be considered (Fig. 14.12). In principle, such a combination is no longer of much use: the crumhorn has a capped reed (Fig. 14.13), which gives the player less control on overblowing. The reliable playing range of the crumhorn is a little bit more than an octave, usually a major ninth. Because of the limited range, music for crumhorns is usually played by a group of instruments of different sizes and hence at different pitches. Such a group is known as a consort of crumhorns. Crumhorns make a strong buzzing sound but are quieter than their conical-bore relatives.

14.3 The Brass Family

14.3.1 The Lip Reed

Now let us apply everything that we know about reeds to brass instruments. What is the reed for the brass family? Is it soft, like for woodwinds, or hard, like for harmonics and accordions?

The role of the reed in instruments of the brass family is played by the lips of a performer. And it is sometimes called the lip reed. The shape of such a "reed" is quite different from the metal strip or piece of cane, but this is not essential. What is important is that the lip reed has a small opening with soft boundaries through which the air is provided to the instrument. Excess pressure inside the mouth pushes outward on the lips, tending to open them with a rolling motion. But when the opening becomes too big, the pressure difference drops, and tension in the lips closes the lips again. The lips can vibrate alone, like other reeds, producing a buzzing sound; positive feedback from the resonating tube can strengthen this sound greatly and give it more stable and musical qualities.

The lip reed by its elastic properties is in between the reed categories of hard and soft. The tension in the lip muscles can cause a sharp enough line in spectrum to excite the instrument exactly in some particular upper mode. The experienced players can excite the mode number seven without "touching" the sixth or eighth, which is a very precise fit in frequencies. Yet for each mode the lip reed is soft enough to be controlled by the feedback from the pipe resonance. The biggest difficulty for a player is sitting in upper modes that are not separated enough from

Fig. 14.14. Tuba.

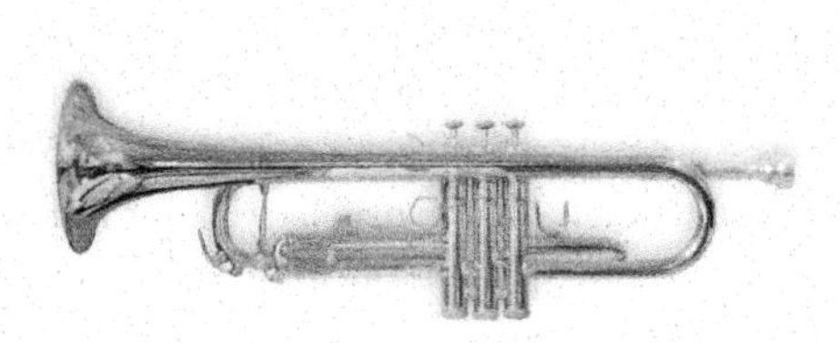

Fig. 14.15a. Trumpet.
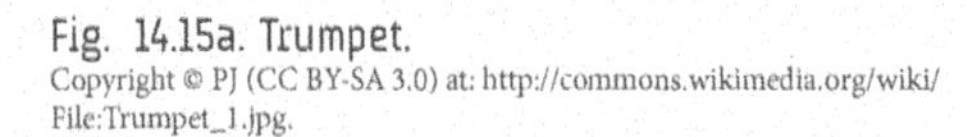

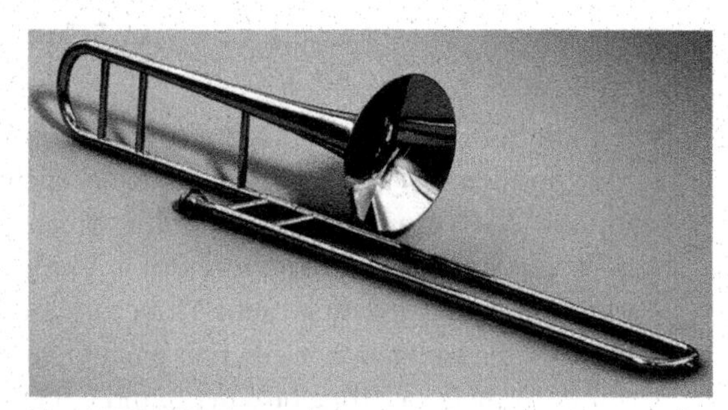

Fig. 14.15b. Trombone.

each other, so there is a danger of jumping back and forth or even sliding continuously from one mode to another.

14.3.2 Some Special Features of the Brass Family

As with other reed instruments, the brass family has a displacement node at the mouthpiece because the blown end of the tube is effectively closed. So, we can expect two subfamilies of brass: the one with cylindrical and another with conical bore. The cornet, alto and baritone horns, and tuba (Fig. 14.14) have conical shapes; meanwhile, the trumpet, trombone, and French horn are cylindrical for the major portion of their length (Figs. 14.15a and b).

In the case of the conical body, the entire bore is already flared. The open end is quite large and needs a little more of a bell to radiate efficiently. The cornet family, also with very small cone angle of the bore, has a relatively smooth and gradual bell. But the trumpet, which looks very similar to the cornet without a bell due to the cylindrical shape of a bore, would have an extremely narrow opening and a soft voice comparable to a flute. So the bell is an important feature for the trombone and trumpet, actively altering the tone color.

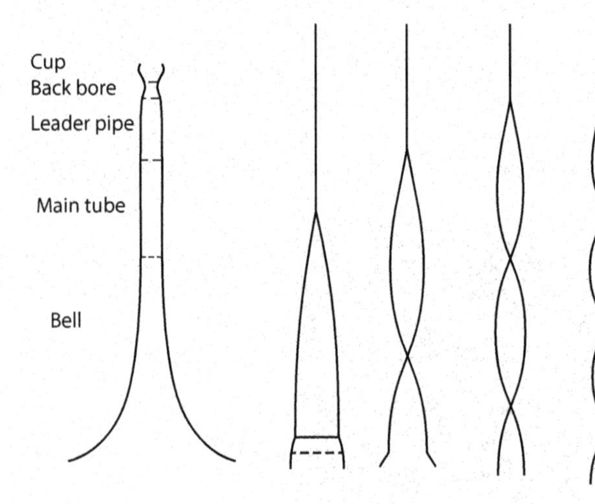

Fig. 14.16

For the trumpet family we may expect the existence of only odd harmonics in the spectrum. Like the case of the clarinet, the overblown instrument should give us the mode 19 semitones higher. But the real trumpet gives a complete harmonic series with a missing fundamental. When we add a mouthpiece and a flare bell to a simple cylindrical pipe, like a garden hose, we are changing the spectrum drastically. Through trial and error, people many years ago arrived at the mouthpiece-and-bell combination, which changed the standing-wave pattern (Fig. 14.16) of overblown notes, including both odd and even harmonics.

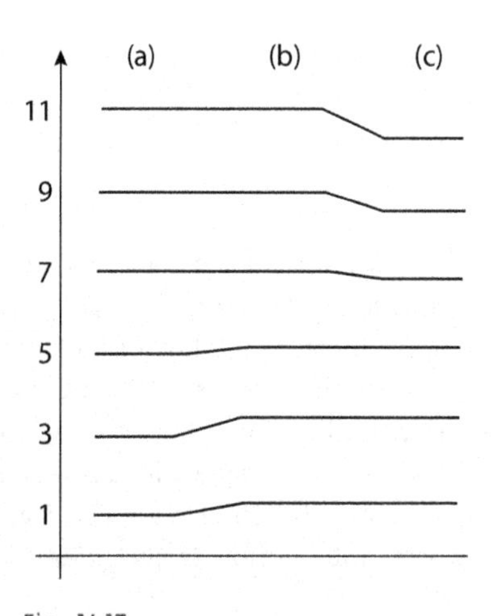

Fig. 14.17
Adapted from Musical Acoustics by Donald Hall, 2002 Cengage Learning, Inc.

How the mouthpiece and the bell influence the final spectrum of the trumpet is shown in Fig. 14.17. We took the "legal" spectrum of a cylindrical pipe with only odd-numbered harmonics, indicated as "tube" in the picture. Adding a bell causes the shift of all lines a little bit upward (the "bell" part in the figure). Notice that the lowest modes are influenced much more than the higher ones. The mouthpiece ("mouthpiece" in the figure) lowers the modes, and the higher modes are lowered more than the lowest. The resulting spectrum has harmonics 2, 3, 4, 5… of a different fundamental, not the one with which we started. The price paid for such kind of alignment of upper modes is obvious: the first mode frequency is entirely different from the fundamental frequency of the harmonic series obtained; this mode is not supported by a series and is not musically useful.

The change of pitch in brass family is reached by using valves, which effectively lengthen the size of a tube. The simplest and most straightforward way to adjust a resonator for a needed frequency is used in the trombone (Fig. 14.15). Note that the positions corresponding to the decreasing of a pitch by a semitone are not equally spaced, because each must represent a 6% increase for the effective length for the preceding position. This unequal spacing has exactly the same reason as the decreasing spacing of guitar frets.

14.4. Radiation

Now it is time to understand how vibration inside an instrument can escape and reach our ears, which means we should discuss the efficiency of reed instruments. There is one interesting difference between woodwinds and the brass family: woodwind instruments are non-directional. This means that the sound produced propagates in all directions with approximately equal volume. Brass instruments, on the other hand, are highly directional, with most of the sound produced traveling straight outward from the bell. This difference makes it significantly more difficult to record a brass instrument accurately. It also plays a major role in some performance situations, such as in marching bands. We already discussed the similar non-uniformity of radiation for violins. This explains why in big orchestras brass players sit in the rear row, pretty much centered, with bells directed toward the audience.

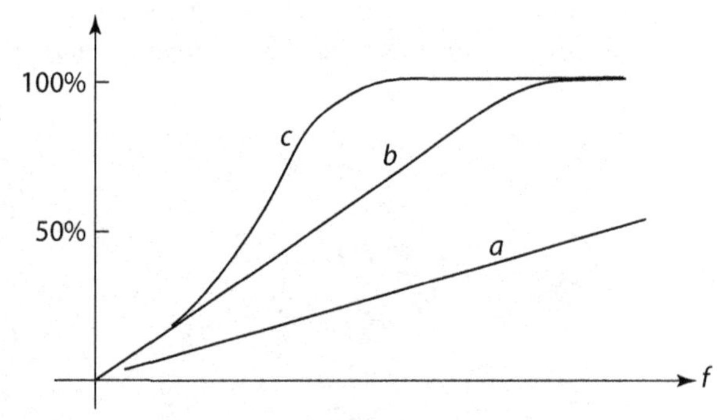

Figs. 14.18a, b, and c.
Adapted from Musical Acoustics by Donald Hall, 2002 Cengage Learning, Inc.

The radiation efficiency is the ratio of acoustic pressure just outside of the end of the instrument to that just inside. If a traveling wave keeps going past the end, it has 100% radiation efficiency; if most of it is reflected at the end and stays inside to build up a standing wave, this means less radiation efficiency.

The radiation efficiency of a small opening (narrow end or a tone hole) is very small for low frequencies. It gradually increases with higher frequencies, finally reaching 100% for wavelengths shorter than half of the hole's diameter (Fig. 14.18a). A large hole lets more sound out and naturally reaches 100% of efficiency earlier (Fig. 14.18b). A trumpet bell keeps the lowest modes back in the cylindrical pipe, but the bell helps the highest modes to escape. The overall behavior of a trumpet with a bell is shown in Fig. 14.18c. The efficiency reaches its maximum at approximately 1600 Hz for a typical trumpet. The leveling-off point for woodwinds is typically around 1600 Hz for clarinet, 1400 Hz for oboe, and 400 Hz for bassoon.

As we may see, the radiation of low frequencies is always much less than the radiation of higher harmonics. As a result, the sound that we hear is brighter and richer than that inside of the instrument.

14.5 Effect of Mute

A **mute** is a device fitted to musical instrument to alter the sound produced by affecting the timbre, reducing the volume, or most commonly both. A variety of mutes have been used on brass instruments, all of which either squeeze inside the bell of the instrument or are hung or clipped to the outside of the bell. These mutes are typically made out of aluminum, brass, or copper. The trumpet with three most common mutes is shown in Fig. 14.19. The mute is another active part of brass instruments because it can drastically change the character of the voice of the considered instrument.

The most common type is the straight mute, a hollow, cone-shaped mute that fits into the bell of the instrument. This results in a more metallic, sometimes nasal sound, and when played at loud

Fig. 14.19. Trumpet with three common mutes.

Fig. 14.20. Wah-wah (Harmon) mute.

volumes, it can result in a very piercing note. Straight mutes have small pieces of cork attached to the end that squeeze against the inside of the bell and hold the mute in place. Straight mutes are available for all brass instruments, including the tuba.

Another of the most common brass mutes is the cup mute. Cup mutes are similar to straight mutes, but attached to the end of the mute's cone is a large lip that forms a cup over the bell. The result is removal of the upper and lower frequencies and a rounder, more muffled tone.

The wah-wah mute (also known by the brand name, Harmon) is a hollow, bulbous mute in two parts. (See Fig. 14.20 for a more detailed view.) Unlike the more common straight or cup mutes, the Harmon mute has a solid ring of cork that completely blocks all of the air leaving the bell and forces the entire air column into the mute. In a hole on the front of the mute, there is a cup on a tube that can be slid in or out or removed completely, depending on the composer's direction or the player's preference. The mute produces a sound perhaps best described as a high-pitched buzz. Harmon mutes are available for many brass instruments but are commonly used only for trumpets and trombones.

Miles Davis often played through a Harmon mute without the stem. This greatly shaped the character of his sound and greatly influenced the jazz community with such classic tracks as "All Blues."

Summary, Terms, and Relations

Bernoulli effect: in the area where the speed of fluid is high and the pressure is low.

Conical bore of the instruments leads to the appearance of even harmonics in spectra.

Mouthpiece, bell, bore, and the mute all influence the spectra of the resulting sound of the woodwinds.

The trumpet demonstrates the spectrum with odd and even harmonics present but with a missing fundamental.

The bell increases radiation efficiency, especially in the low-frequency range.

The mute has its own spectrum; it not only decreases the amplitude of the resulting sound but also shapes the spectrum of the instrument.

Questions and Exercises

1. Discuss a change of spectrum of a tube closed at one end when the shape of this pipe is changed from cylindrical to conical.

2. Discuss different examples of the Bernoulli Effect in everyday life.

3. Discuss the features of the clarinet, saxophone, and oboe. What are their general features? How do they differ?

4. Why does a flute of the same size as a clarinet have a voice that is approximately one octave higher?

6. The clarinet's lowest note is D_3. What is the length of an idealized closed pipe that has this frequency for the fundamental mode? Discuss why the actual length of the instrument is bigger.

5. The oboe's lowest fundamental frequency is 233 Hz. What is the length of an idealized closed pipe that has this frequency for its fundamental mode? Discuss why the actual length of the instrument is shorter.

7. By what fraction should you move a trombone's slider to raise the desired pitch by two semitones?

8. What is the main specific feature of the trumpet's spectrum?

Chapter 15

The Human Voice

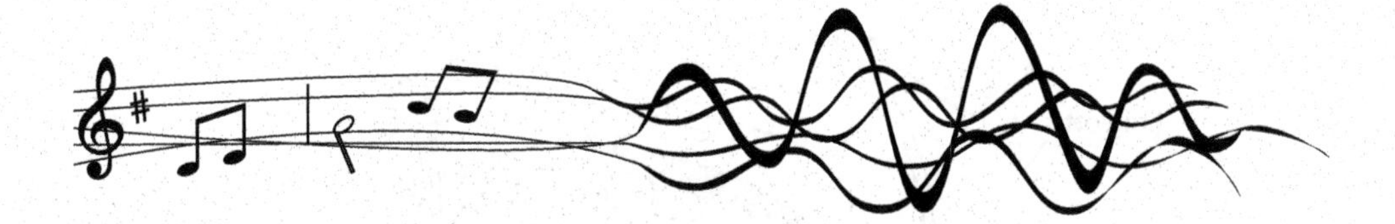

15.1 A Natural Instrument

Now it is time to consider a musical instrument that nearly everybody has in his/her possession: the human voice. The closest relative of it out of the all families of instruments is, definitely, the woodwind family. Like any wind instrument, the human voice has

- an air reservoir, which maintains pressure above atmospheric (our lungs); ;
- an outlet channel with a small constriction, where the airflow can be modulated and even interrupted (the larynx);
- and a resonator cavity to strengthen and support the sound created in the outlet channel (the throat and mouth, together with everything that we have in it: teeth, tongue, lips, etc).

When we take a breath, air is pumped down into the lungs by raising the rib cage. When we breathe out, abdominal muscles contract to push it out. An adult person holds approximately 4 liters of air in his/her lungs, though can hold up to 6 liters with a very deep breath. We do not use all this air while speaking, singing, or breathing; only about half a liter of air moves in and out with each breath.*

The **trachea** is a long tube that connects the lungs to the **vocal tract**, which consists of the throat, mouth and nose (Fig. 15.1). The **larynx** is a hollow piece of cartilage that sits at the top of the trachea and esophagus (which runs alongside the trachea and connects the throat to the stomach), approximately 8 cm long and 4-5 cm wide (Fig. 15.2). A flap-like valve called the

* Hall, *Musical Acoustics*, p. 193

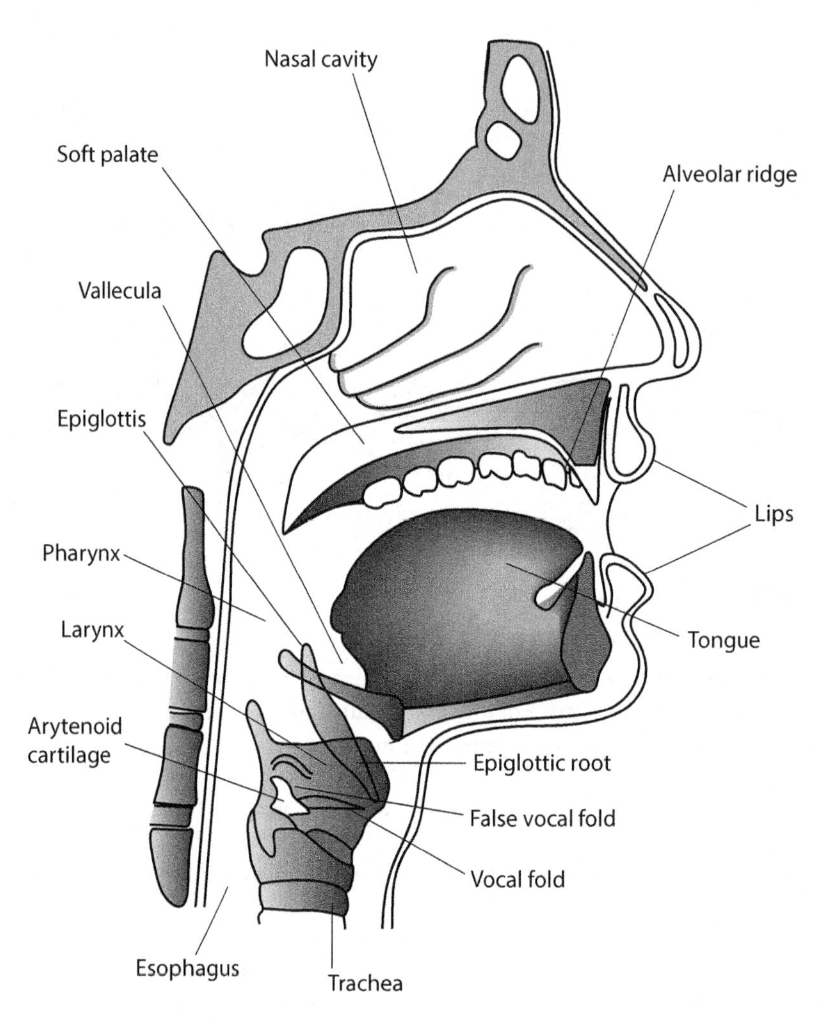

Fig. 15.1. Vocal apparatus.

epiglottis sits atop the larynx and drops down when eating and swallowing to prevent food and water from entering the trachea. Remember the saying, don't talk while you eat? The epiglottis remains open for phonation; you risk choking if you talk and eat!

At the top of the trachea is a hollow boxlike structure of cartilage called the **larynx** (Fig. 15.2). The larynx, approximately 8 cm long and 4–5 cm wide, also joins the trachea and esophagus. The epiglottis is a flap-like valve on the top of the larynx that drops down during swallowing to prevent pieces of food from entering the trachea, but it is open for phonation. Do you remember how your parents were urging you: Don't talk while you are eating?

However, the vocal cords (or vocal folds), two soft ridges of tissue on the inside of the larynx, serve as a back up to the epiglottis and can also block the trachea. They, too, close when you swallow and open when you breathe.

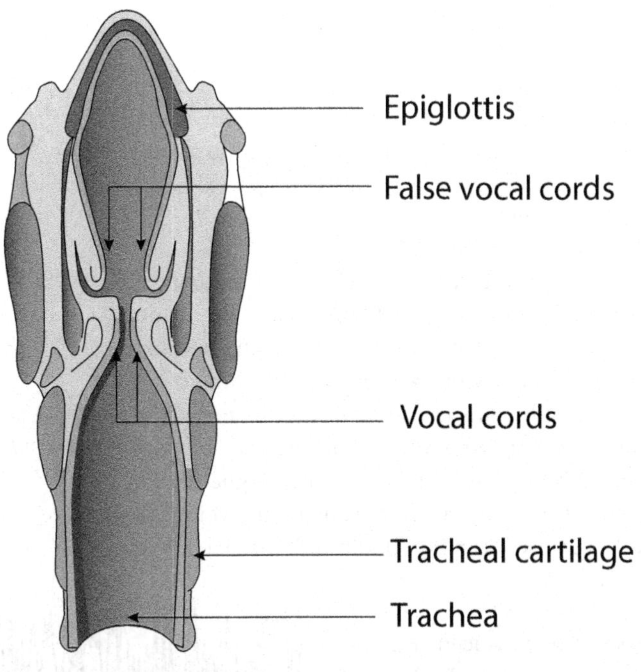

Fig. 15.2. Larynx in detail.

Above the larynx is the throat cavity, which opens to the outside via the mouth. The throat may not open into the nose depending on the position of the soft palate. The tongue, lops and teeth participate in blocking and restricting airflow.

15.2 Producing Sound

To produce sound the vocal cords close and the lungs apply pressure (approximately .03 atm for speech and .1 atm for singing†). This pressure opens the vocal cords, sending spurts of air into the vocal tract. The vocal cords are moved by small muscles in a V-shape which stay together in the front and widen at the back. The opening between the two cords, the top of the V, is called the glottis. It is approximately 2 cm long and 1 cm wide when open. The tension of the muscles while air is passing through causes the cords to vibrate at a frequency of 70-200 Hz for a male and 140-400 Hz for a female.‡ Males have longer, larger vocal cords, which accounts for this difference; the range also varies up or down an octave while singing. The range is 80-1100 Hz (that is E2 to C6) for males and females.

† Ibid
‡ Ibid

15.2.1 Consonants

No two vocal tracts are alike – the size and shape differs from person to person creating different sounds. Each unique sound is called a **phoneme**. For acoustical purposes it is good to classify the phonemes into five groups: plosives, fricatives, approximants, pure vowels, and diphthongs.

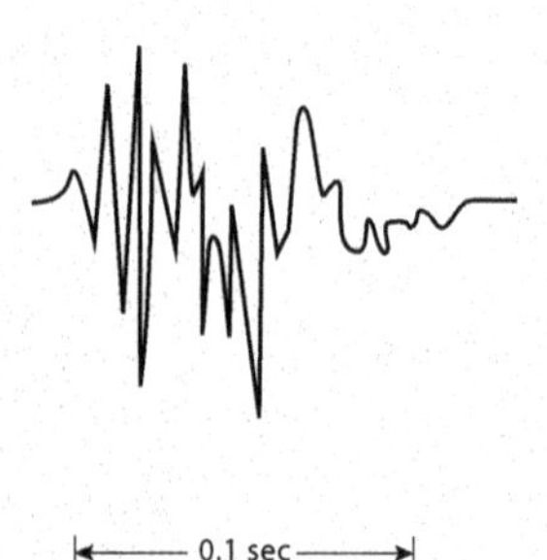

Fig. 15.3a. Typical waveform of sound while producing a plosive.
Adapted from Musical Acoustics by Donald Hall, 2002 Cengage Learning, Inc.

Plosive consonants are produced by completely blocking the vocal tract, opening it for only one short burst of air. You cannot sing or hold the plosive; it is a transient sound lasting no more than 0.1 s (Fig. 15.3a), though the blocking and opening of the vocal tract can occur at the lips (pushing air out to create a "p" sound, or adding vibrations and voice to create the "b"), soft palate (the "k" or "d" with voice), or front of the tongue (the "t" or "g" with voice). Try and hold a "p" without adding voice (the "eee") – you can't! Because it is nonperiodic and nonstead, there is no definite pitch.

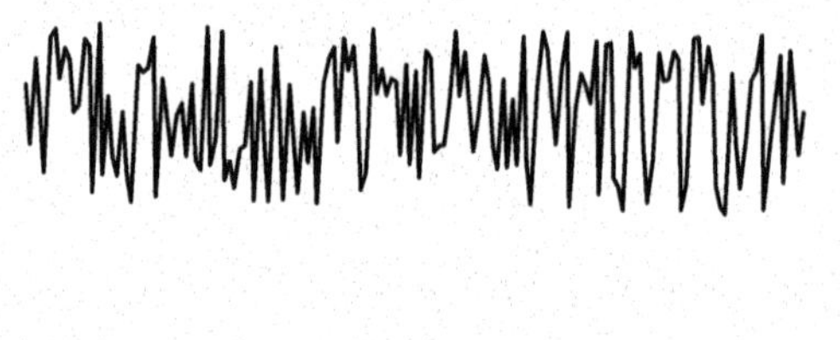

Fig. 15.3b. Typical waveform of sound while producing a fricative.
Adapted from Musical Acoustics by Donald Hall, 2002 Cengage Learning, Inc.

The fricatives also come in voiced and non-voiced pairs: f and v, th (as in thin) and th (as in this), s and z, sh and zh. The only unvoiced fricative that does not have a pair is h (home). The fricatives can be held or sustained steadily for any length of time similar to white noise (Fig. 15.3b); a continuous range of frequencies with no definite pitch. This sound is created by air moving through a small opening faster than critical speed. As with plosives, the opening can be formed from several places in the vocal tract: tongue and palate (s/z and sh/zh), the lips and teeth (f, v, and th), and glottis (h). The frequency varies slightly (around 4–6 KHz and sh, around 3 KHz) but there is no distinct frequency within the range. The spectrum is non-harmonic, and the production of these consonants does not involve the vibration of the vocal cords.

The fricatives also come in voiced and non-voiced pairs: f and v, th (as in thin) and th (as in this), s and z, sh and zh. The only unvoiced fricative that does not have a pair is h (home). The fricatives can be held or sustained steadily for any length of time similar to white noise (Fig. 15.3b); a continuous range of frequencies with no definite pitch. This sound is created by air moving through a small opening faster than critical speed. As with plosives, the opening can be formed from several places in the vocal tract: tongue and palate (s/z and sh/zh), the lips and teeth (f, v, and th), and glottis (h). The frequency varies slightly (around 4–6 KHz and sh, around 3 KHz[§]) but there is no distinct frequency within the range. The spectrum is non-harmonic, and the production of these consonants does not involve the vibration of the vocal cords.

§ Hall, *Musical Acoustics*, p. 194

The approximants include several different groups of sounds between fricatives and vowels, including:

- the semivowels w (like in why) and j (or "y" like in yes). Semivowels, by definition, contrast with vowels by being non-syllabic. According to the standard definitions, semivowels (such as [j]) contrast with fricatives in that fricatives produce turbulence, while semivowels do not;

- the liquid consonants lateral l (like in led) and rhotic r (like in red). Elsewhere in the world, two liquids of the types mentioned above remain the most common attribute of a language's consonant inventory except in North America and Australia. Most indigenous Australian languages are very rich in liquids, with some having as many as seven distinct liquids. On the other hand, there are many indigenous languages in South America and eastern North America, as well as a few in Asia and Africa, with no liquids at all. Polynesian languages typically have only one liquid, which may be either a lateral or a rhotic;

- the nasal consonants (represented in English by "nasal stops" m, n, ng) are created by airflow that is directed through the nose. These consonants encompass the vowel–consonant transition. While phonating a nasal, the vocal cords already are vibrating. The nose in open, which allows a steady voiced sound, such as humming. That sound can be held indefinitely, and it has clear pitch and a harmonic-series spectrum. It can be called a nasal vowel as far as acoustics is concerned. From the side of linguistics, there is a difference: nasal consonants do not create syllables. The nasals differ from each other only in the place of articulation of the final consonant: lips for m, tongue for n, and soft palate for ng.

15.2.2 Vowels: Pitch

The vowels are steady, voiced sounds with definite pitch. In phonetics, a **vowel** is defined as a sound in spoken language, pronounced with an open vocal tract so that there is no build up of air pressure at any point above the glottis. The waveforms of vowels are periodic, and their spectra are harmonic series.

The production of a vowel involves the steady vibrations of vocal cords. When a stream of air is sent between almost-closed vocal cords, they start to vibrate because of the Bernoulli effect, which was considered in Chapter 14. The frequency of these vibrations completely defines the pitch of our spoken or sung vowel.

So, **regardless of what vowel may be involved, the pitch of the voice is completely determined by the tension of the vocal cords.** Trained singers can sing pitches in a musical scale accurately and on demand.

A diphthong refers to two adjacent vowel sounds occurring within the same syllable. Technically, a diphthong is a vowel with two different targets: that is, the tongue moves during the pronunciation of the vowel. In most dialects of English, the words *eye*, *hay*, *boy*, *low*, and *cow* contain diphthongs. The essence of a diphthong is transition from one sound to another. Diphthongs

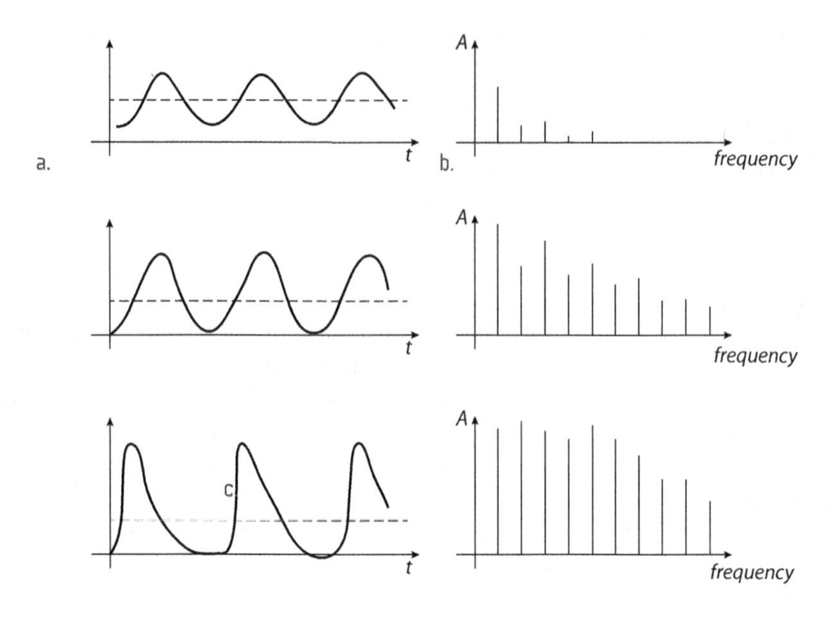

Figs. 15.4.a, b, and c. (a) Motion of vocal cords while speaking a vowel softly. Upper harmonics are suppressed. Adapted from: Musical Acoustics by Donald Hall, 2002 Cengage Learning, Inc., after Strong and Putnik. (b) Motion of vocal cords at moderate intensity: spectrum shows appearance of some upper harmonics. (c) Motion of vocal cords while speaking loudly: spectrum shows the whole set of upper harmonics.
Adapted from Musical Acoustics by Donald Hall, 2002 Cengage Learning, Inc.

are extremely difficult for singers because on long notes there is a problem of how much to lean toward the transition and how soon finally to go through it.

15.2.3 Vowels: Loudness

Vowels are of special concern to musicians because they have definite musical pitch and can be sustained for the assigned length of each musical note.

We already know that the particular pitch of a vowel is created by the periodic vibration of the vocal cords. The timbre of the sound right after the vocal cords does not have anything even close to a human voice which we can expect. It is just a buzzing sound, the result of series of puffs of air, whose exact nature depends on how forcefully the air is going through the larynx.

Fig 15.4a shows the motion of vocal cords while a person is whispering. As we may see, the vocal cords practically do not block the airflow, oscillating in an almost sinusoidal manner. The spectrum of these oscillations demonstrates a very small number of upper harmonics, maybe 2 or 3, significantly suppressed.

When a person speaks a vowel of some moderate intensity (Fig. 15.4b), the vocal cords come closer and closer to each other, almost blocking the airflow at times. The waveform of such a wave is not close to sinusoidal anymore, which is demonstrated in the corresponding spectrum.

If the loudness is increased even more, the vocal cords block the glottis for approximately one third of each period, which results in an even larger numbers of upper harmonics. The form of a signal is now far from sinusoidal. The lowest 5–6 harmonics have comparable amplitudes (Fig. 15.4c), and the spectrum drops after them.

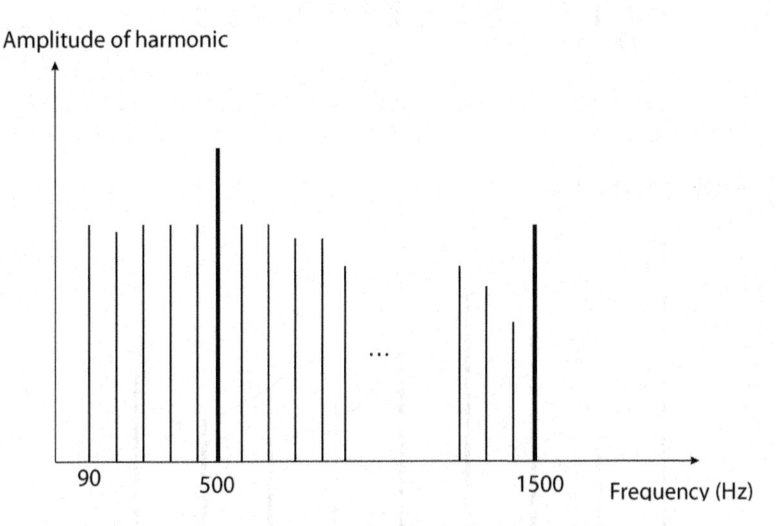

Fig. 15.5a. Iron Man tries to say a vowel with pitch 90 Hz, but his vocal tract has resonating line at 500 Hz and 1500 Hz: nothing can be heard.

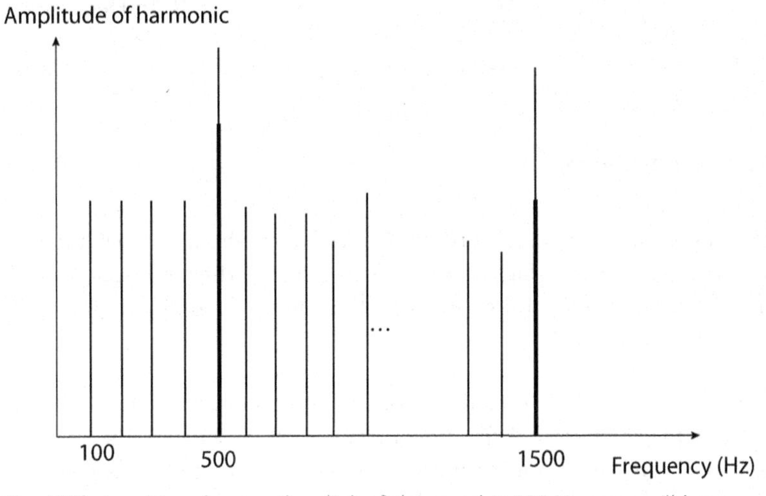

Fig. 15.5b. Iron Man changes the pitch of the vowel to 100 Hz: now we'll hear a very loud sound.

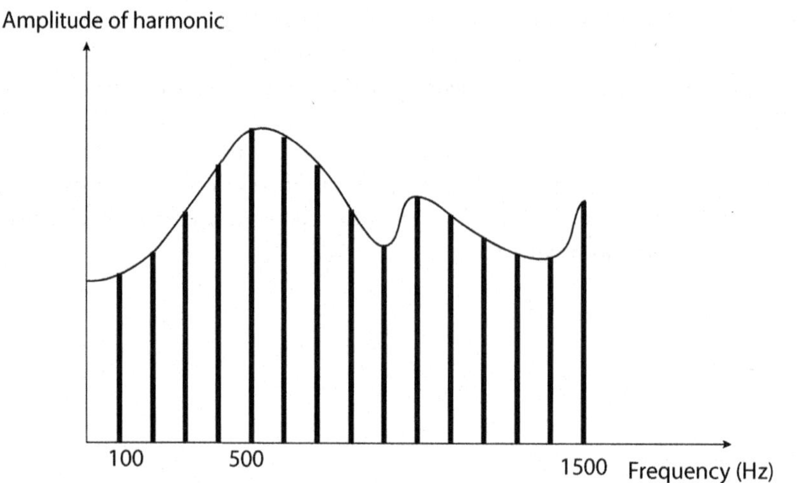

Fig. 15.5c. Resonating peaks of our vocal tract are broadened and weakened, so all frequencies are enhanced. Here the person is speaking a vowel of 100 Hz pitch.

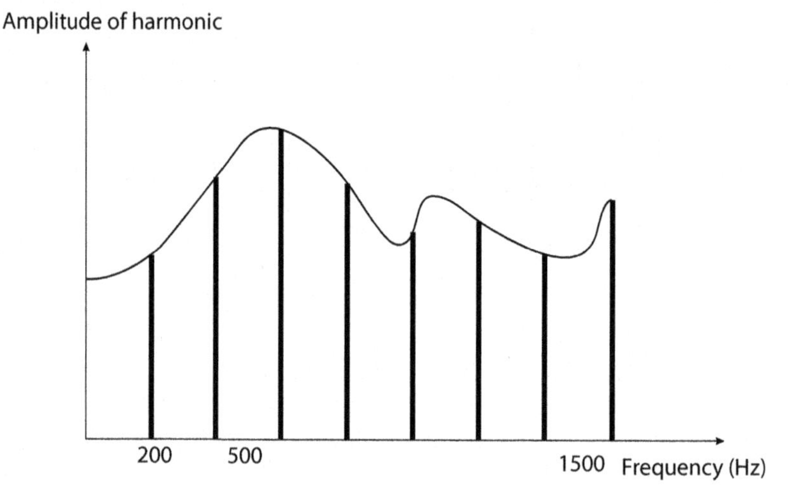

Fig. 15.5d. A person is speaking the same vowel as in Fig. 15.5c but one octave higher, at 200 Hz pitch.

15.3 Vowels: Formants

While speaking a vowel, our vocal cords vibrate with some frequency, creating the pitch of the resulting sound. But this is so far just a sound, not words, which we usually use for singing or exchanging information. How do we organize this uniform sound into human speech? Let us consider this step by step.

What is our vocal tract? It is a resonator, which supports some frequencies and suppresses others. This process is completely defined by a form of this resonator. The vocal tract is a tube, approximately 17 cm long, closed from the inner end, as for all reed instruments, and open at the mouth. For the first rough approximation, we consider our vocal tract as a uniform cylinder.

The spectrum of a cylinder closed from one end has only odd multiples of v/4L, or for given length 500, 1500, 2500, 3500 Hz, …

If we were Iron Men, the spectrum of our vocal tract would look like one in Fig. 15.5a. Here the thick lines correspond to the frequencies of our resonator's cylindrical tube, and fine lines correspond to the oscillations of the vocal cords with frequency 90 Hz (from Fig. 15.4c). No one line delivered by the vibration of the vocal cords into the vocal tract is supported. As a result, not much sound, if any, would come from the mouth of Iron Man. The situation will change drastically if the frequency of the vibration of the vocal cords falls into the line of the vocal tract (Fig. 15.5b), for instance, at 100 Hz. Iron Man suddenly starts to speak.

We are not Iron Men, which is good. All the walls and other details of our "pipe" are made of soft tissue, so all the lines of the vocal tract resonances are broadened and weakened (Fig. 15.5c).

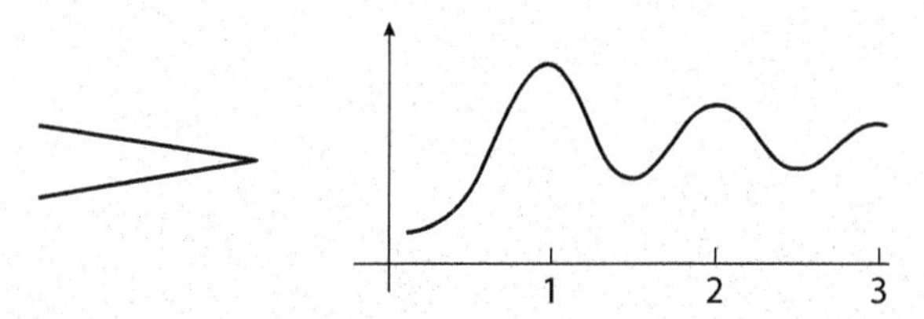

Fig. 15.6a. The conical shape of a pipe slightly corresponds to the sound "a" as in "had."
Adapted from Musical Acoustics by Donald Hall, 2002 Cengage Learning, Inc., after Strong and Putnik

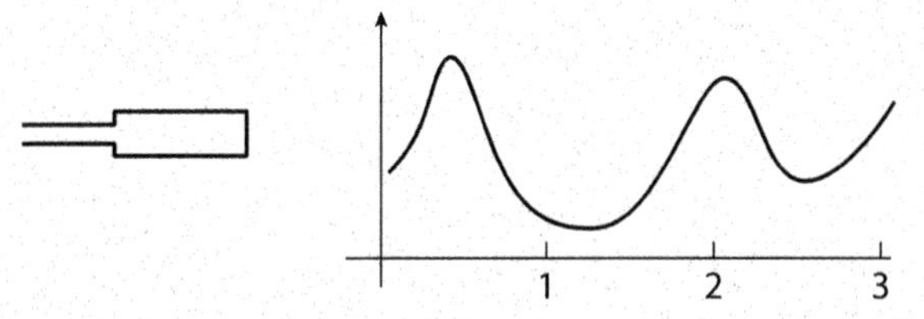

Fig. 15.6b. This shape of a pipe slightly corresponds to the sound "ee" as in "heed."
Adapted from Musical Acoustics by Donald Hall, 2002 Cengage Learning, Inc., after Strong and Putnik

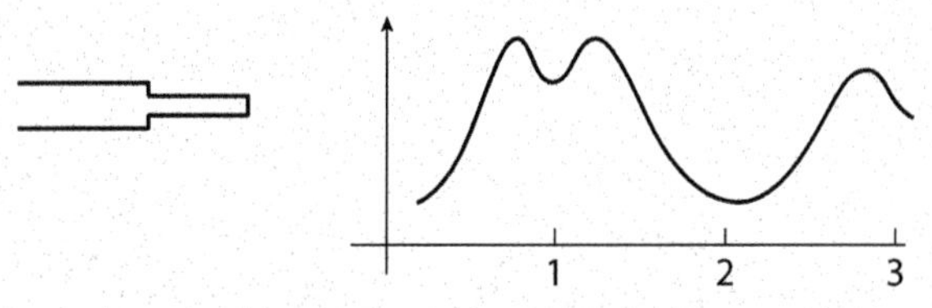

Fig. 15.6c. This shape of a pipe slightly corresponds to the sound "au" as in "bought."
Adapted from Musical Acoustics by Donald Hall, 2002 Cengage Learning, Inc., after Strong and Putnik

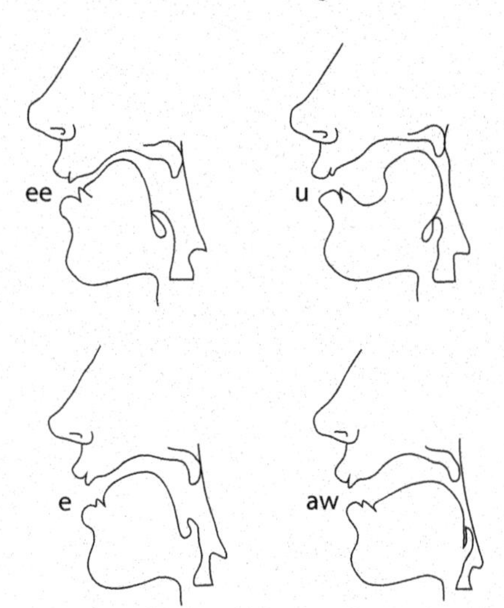

Fig. 15.7a. The actual configuration of the vocal cord for different vowels.
Source: Musical Acoustics by Donald Hall, 2002 Cengage Learning, Inc.

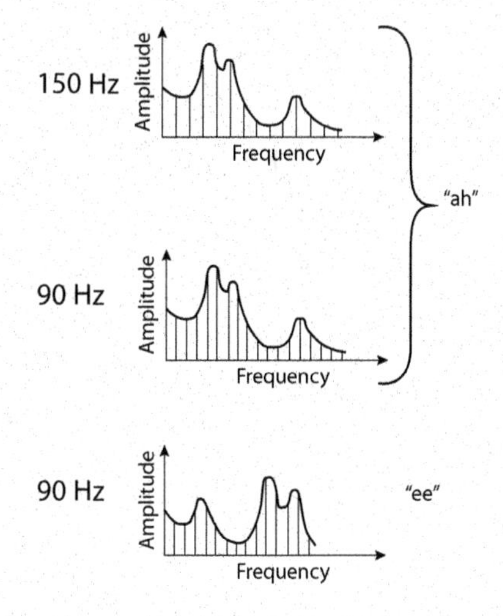

Figs. 15.7b, c and d. (b) The sound "ah" spoken with a pitch of 150 Hz. (c) The sound "ah" spoken with a pitch of 90 Hz. (d) The sound "ee" spoken with a pitch of 90 Hz.
Source: Musical Acoustics by Donald Hall, 2002 Cengage Learning, Inc.

All spectral components remain in the radiated sound, and the vicinity of each resonance provides a mild boost of all frequencies. The full spectrum of upper harmonics with the fundamental pitch 100 Hz is produced. Each such frequency range, in which the amplitudes of spectral components are enhanced, is called a **formant**.

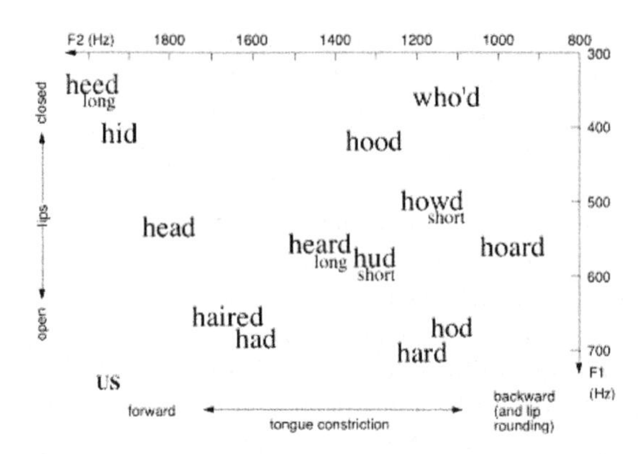

Fig. 15.8. Position of first (horizontal axis) and second (vertical axis) formants for different spoken sounds.
Source: http://newt.phys.unsw.edu.au/jw/voice.html

Now suppose that a person sings one octave higher than in Fig 15.5c, which is 200 Hz, keeping the vocal tract in the same shape as the form of the cylindrical tube. Then the harmonic series of vocal cords contains all multiples of a 200 Hz fundamental, but the formants remain the same (Fig. 15.5d). The sound represented in Fig. 15.5c and d will have similar character and will represent the same vowel quality. But the height of the pitches will differ by an octave.

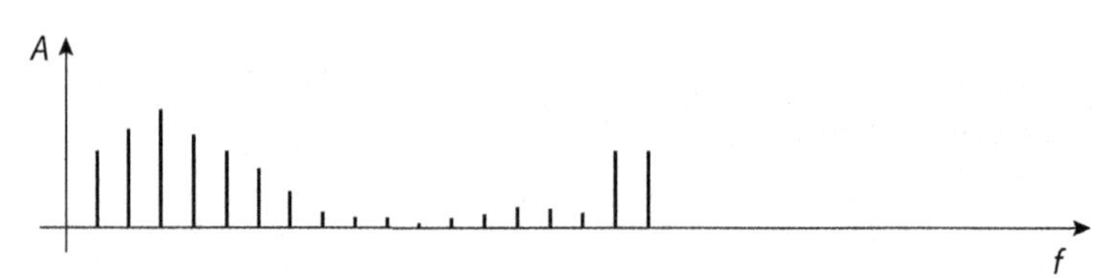

Fig. 15.9a. Spectrum of sound spoken softly. Second formant is not filled in.
Source: Musical Acoustics by Donald Hall, 2002 Cengage Learning, Inc.

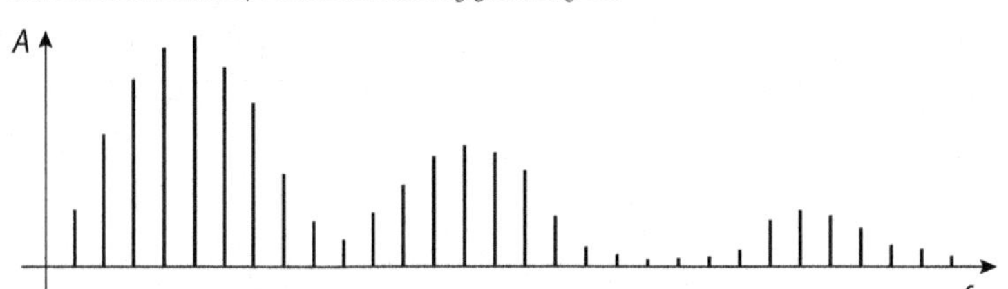

Fig. 15.9b. Spectrum of a sound spoken louder. Note the appearance of the second formant.
Source: Musical Acoustics by Donald Hall, 2002 Cengage Learning, Inc.

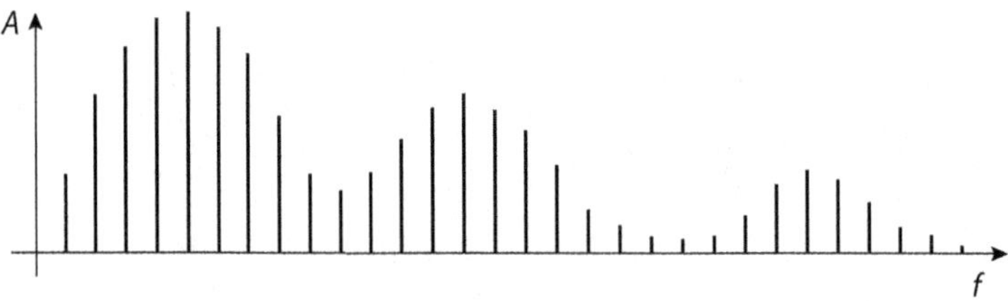

Fig. 15.9c. Spectrum of sound spoken louder than in Fig. 15.9b. The upper harmonics are strong enough to fill the second and part of the third formant.
Source: Musical Acoustics by Donald Hall, 2002 Cengage Learning, Inc.

What should we do to speak another vowel? From all considered above, we already are able to make a conclusion: we need to change the configuration of a vocal tract, change the form of our resonator. Figs. 15.6a, b, and c show several other forms of possible resonators, whose frequencies can be calculated involving more mathematics: (a) a cone of length 17 cm; (b) a cylindrical bottle of length 9 cm and cross-sectional area 8 cm^2 with a neck of length 6 cm and cross-sectional area 1 cm^2; and (c) a narrow tube (length 8 cm, area 1 cm^2) opening into a wider reservoir of length 9 cm and area 9 cm^2. These calculations have been made in the book "Music, Speech, Audio" by W. Strong and G. Plitnik.

The case shown in Fig. 15.6a shows formants around 1, 2, 3, … KHz, giving the sound close to "*a*" in "had." But our vocal tract definitely is far from a shape of a cone. There are other sophisticated shapes, whose first formants fall into similar frequencies. Fig. 15.6 b roughly corresponds to the vocal tract shape with resulting formats for "*ee*" like in "heed" and 15.6c represents "*aw*" like in "bought." The real configuration is shown in Fig. 15.7a (from the "Speech Chain," by Peter B. Denes and Elliot N. Pinson, 1963).

Very useful illustrations of the format spectra are shown in Figs. 15.7b and c: the sound "ah" for fundamental 150 Hz (b) and 90 Hz (c). Note that the form of formants stays the same, only the spacing between harmonics decreases. The sound "ee" with pitch 90 Hz is shown in part (d) of this figure. Spacing between harmonics is the same as in part (b), but the configuration of the formants is different.

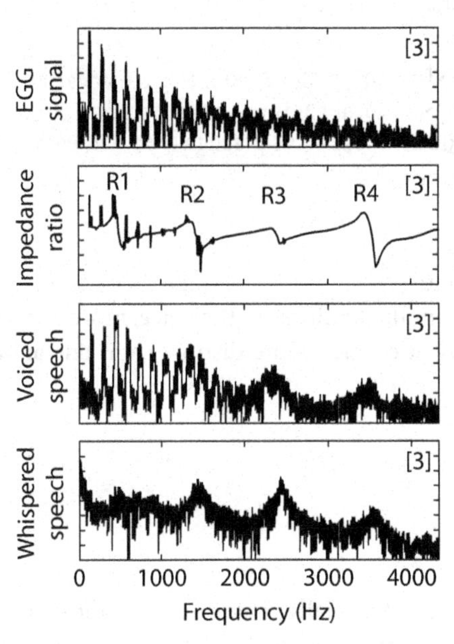

Fig. 15.10a. Position of harmonics relative to formant peaks for "heard." Note that the lines corresponding to the pitch fall in between formants.
Source: http://newt.phys.unsw.edu.au/jw/voice.html

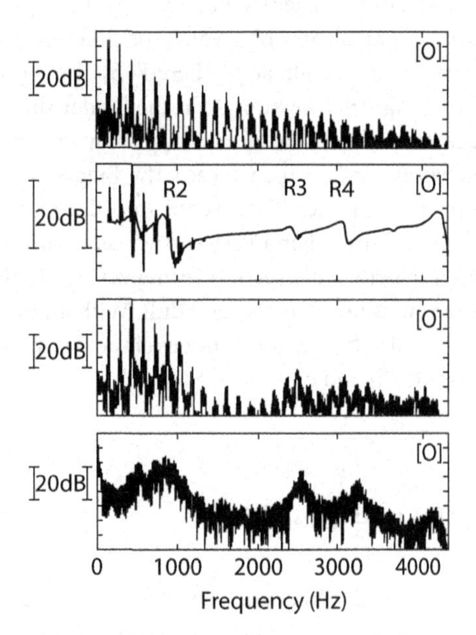

Fig. 15.10b. Position of harmonics relative to formant peaks for "hot." Note that the lines corresponding to the pitch are supported by formants.
Source: http://newt.phys.unsw.edu.au/jw/voice.html

15.4 Spoken Vowels

Although we actually are aided in vowel recognition by third and fourth formants, the most important for the recognition of a vowel is the position of the first and second formants. Let us discuss Fig. 15.8. The horizontal axis represents the first formant frequency, and the vertical axis represents the second formant frequency so that each point of a diagram corresponds to a pair of formants. Regions are shown on the graph instead of points because there is a considerable variability not only from one speaker to another but even from time to time in each person. As a result, there is a whole range of formant-frequency pairs. Values of frequencies for men correspond to the lower left end of each region, for children to the upper right, for women intermediate. So, the first formant around 300 Hz and second at 900 Hz are giving us "oo" as in "soon." Some ranges overlap although it is not so obvious from the figure. The overlapping means that the same sound can be perceived in two different ways. Usually only one of two makes sense in the context of the conversation.

There is an elongated area for each vowel. The reason for this is that men, women, and children do not have the same size vocal tracts. So, each person is not able to produce the formants at exactly the same frequency; for instance, a formant pair 700 Hz and 1000 Hz might be perceived as "sought" in a child's speech but "sot" in a man's.

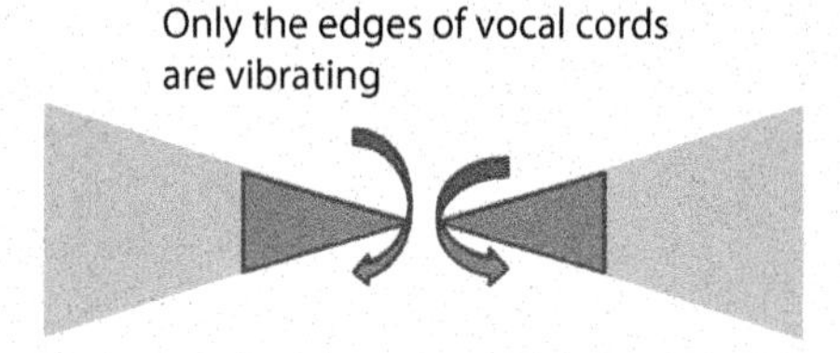

Fig. 15.11. Falsetto: only the edges of vocal cords are vibrating. As a result, vocal cords practically do not block the airflow.

Now we can answer why a very soft voice, actually a whisper, does not sound clear to us. We can hear the sound, but the words are indistinguishable. To understand this, let us analyze Figs. 15.9.a, b, and c. This is a spectrum of a voice for different loudness levels, which is already "filtered" by the vocal tract. From Fig. 15.4 we know that the louder the voice is, the more upper harmonics appear in the spectrum of vocal cords, which means the bigger number of upper harmonics is delivered into the resonator of the vocal tract. For a very soft voice, harmonics disappear before the second formant (Fig. 15.9a), and the vowels are colorless and dull. With an increase in the loudness of the voice, more and more harmonics fill the formant envelope, so the vowels are much more distinct from one another (Fig. 15.9b and c).

15.5 Sung Vowels

Studies of the frequency spectrum of trained singers, especially male singers, indicate a clear formant around 3000 Hz that is absent in speech or in the spectra of untrained singers. It is associated with one or more of the higher resonances of the vocal tract. The additional resonance at 3000 Hz allows singers to be heard and understood over an orchestra, which peaks at much lower frequencies of around 500 Hz.

High sopranos meet the unavoidable problem when the highest pitches are sung. With an increase of the fundamental frequency, the distance between harmonics produced by vocal cords increases, even to a point where a given formant may not have a resonating peak, and fundamental frequency falls in between two formants (Fig. 15.10a). In this situation the voice, being unsupported by resonances of the vocal tract, is losing its richness and depth. The only possibility for a singer to perform these notes is the deliberately shift of formants, as shown in Fig. 15.10b. The vowel accuracy is not held, but the voice sounds musically rich and full. Fig. 15.10 illustrates the preference of "could" versus "cool."

15.6 Some Special Examples of the Human Voice

15.6.1. Falsetto

Falsetto voice is a type of vocal phonation that enables the singer to sing notes beyond the range of a normal voice. The sound of falsetto voice is artificial and even not exactly human. Usually it is one octave higher than the normal range of a particular singer. The strange color of falsetto occurs because of the small number of upper harmonics in its spectrum.

While singing falsetto, only the edges of the vocal cords are moving, while leaving each cord's body relatively relaxed (Fig. 15.11). As a result, the vocal cords practically do not block the airflow (situation, close to Fig. 15.4a), moving almost in a sinusoidal manner, producing just a few upper harmonics. The high pitch of falsetto is a result of the motion of the very edges of the vocal cords, which are much lighter than the base.

15.6.2 Throat Singing

Throat singing, also known as overtone singing or overtone chanting, is a type of singing in which the singer manipulates the formants in such a way that the vocal tract's configuration strictly supports only some of upper harmonics produced by the vocal cords. In this technique, one of the vocal tract resonances is made much stronger, while all the others are weakened. The strong resonance can be made so strong that it selects one of the harmonics and makes it so much stronger that it is perceived as a separate note. So, throat singing can create a sensation that a singer is performing two pitches at the same time. The "strong" harmonic may be pretty far from fundamental: Tuvan throat singers use the 6th, 8th, or even 12th .

Summary, Terms, and Relations

For normal speech the frequencies of the human voice normally extend over the range of 70 to 200 Hz for a man's voice or 140 to 400 Hz for a woman's.

Almost all consonants have no harmonic spectrum.

All vowels have a harmonic spectrum created by the vibration of vocal cords. The vowels have two characteristics: height of pitch, for which the vibration of vocal cords is responsible, and articulation, which is a result of configuration of the vocal tract.

The formant is the enhancement of the amplitude of harmonics around the preferred resonant frequency of the vocal tract.

Questions and Exercises

1. Discuss what the main difference is in the spectra of plosives, fricatives, and vowels.
2. What is the main difference in the spectrum of whispering and speaking loudly?
3. Suppose you speak the word "soon" and then the word "seat" at the same pitch, say 200 Hz. What is the difference in the spectra of these two words? Sketch these spectra using Fig. 15.8.
4. Suppose you sing some vowel at the pitch 440 Hz. The formants corresponding to this vowel are 500 Hz and 2500 Hz. What would the position of the formants be if you sing the same vowel but at a pitch if 300 Hz?
5. To get solid C_6, what vowel should a soprano pronounce?
6. For the first and second formants at 400 Hz and 100 Hz, what two vowel interpretations are possible?
7. Using Fig. 15.9, discuss the problems of speaking in big auditoriums. Should you literally speak more energetically or can you just turn up the gain of the microphone amplifier?
8. Draw the spectra for the vocal cords' vibrations of 200 Hz sent into two cylindrical tracts, one 18 cm long and another only 12 cm long. What should we change in our picture if the vocal cords oscillate at frequency 100 Hz?

Chapter 16

Room Acoustics

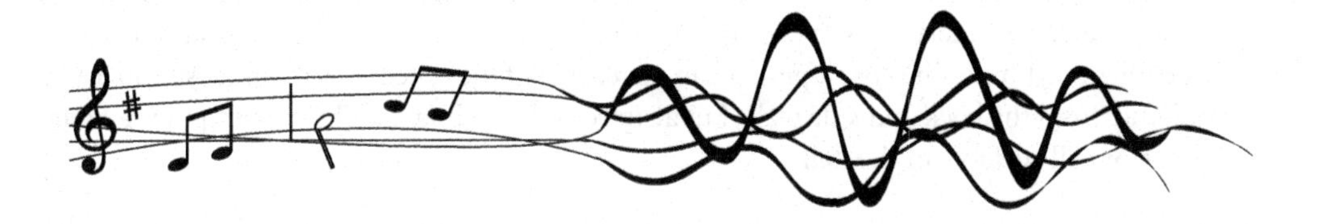

16.1 What Criteria to Choose?

There are several general requirements that should be satisfied for any room serving for the purposes of musical and theatrical performances. These requirements should help not only the listeners but also the performers.

1. **Envelopment**: the listener should be surrounded by sound from all sides, yet the sound should clearly come from the side of the stage, but not from loudspeakers or other sources of sound.

2. **Uniformity**: listeners in all parts of the room should hear as nearly the same sound as possible; especially as concerns the elimination of "dead spots."

3. **Clarity**: each note and word should arrive clearly and undistorted. This is especially important for rooms that are used for speech, such as big lecture halls.

4. Close to the previous requirement is **freedom from echo**. The walls, ceilings, floors, and all objects in the hall reflect sound, but all these reflections should reach the ears of listeners smoothly, not in the form of separate echoes. As we will see this is mostly a requirement for early reflections.

5. The music contains passages of different loudnesses, some of which are very soft. These soft passages should not be disturbed by the sound of the traffic outside, by conversations behind the doors, or by the noise of the ventilation system. As a result, we come to the requirement of **freedom from noise**. We already discussed different requirements for noise levels in Chapter 6. There are the usual ways to struggle with

background noise: build double doors to the concert hall, build thick substantial walls, and use quiet systems for ventilation.

6. **Reverberation** is the persistence of sound in a particular hall after the original sound is produced. A reverberation is created when a sound is produced in an enclosed space, causing a large number of echoes to build up and then slowly decay as the sound is absorbed by the walls and air. This is most noticeable when the sound source stops but the reflections continue, decreasing in amplitude until they can no longer be heard.

How do we meet these acoustical requirements? Each listener should have the possibility of unobstructed sound as uniform as possible for all parts of our hall. As a result, to minimize these differences in distances to any listener, the audience often sits on a slope and in sets of balconies and boxes. The envelopment can be reached if the reflections of direct sound arrive not only from the back and side walls but also from the ceiling and floor or, to put it a better way, from all surfaces in the hall.

A very important requirement for concert halls is the **performer's satisfaction**. For this, the rear wall should not return a strong echo to the stage. A good stage usually has a shell-type shape to smoothly mix all the reflections and to allow all the members of the performing group to hear one another. To avoid trapping an annoying echo between them, the stage should not have hard parallel side walls. Besides annoying the performers, a trapped echo also decreases the loudness for the audience.

16.2 Direct, Early, and Reverberant Sound

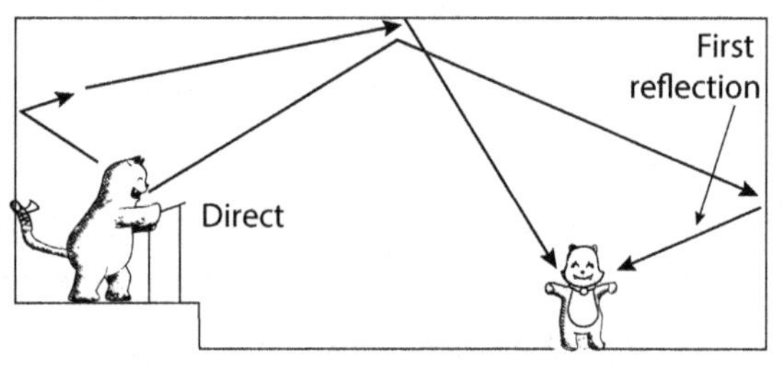

Fig. 16.1. Sound arrives to the listener not only in direct path, but also reflecting from all surfaces.
Adapted from Springer Handbook of Acoustics by Thomas Rossing, 2007 Springer

It is useful to examine how sound travels to the listener in the classroom. The typical length of a classroom is about 20 m, and a large one is about 30–35 m. As a result, the direct sound can reach the listener after a time 0.01 to 0.1 s, depending on the distance of the source from the listener. A short time later, the same sound will reach the listener from one of reflecting surfaces, mostly the walls and ceiling. In Fig. 16.1 these reflections are shown with various times of delays t_1, t_2, t_3. ... The first group of reflections that reach the listener within 50 to 80 milliseconds is called **early reflections,** or just **early sound** (Fig. 16.2).

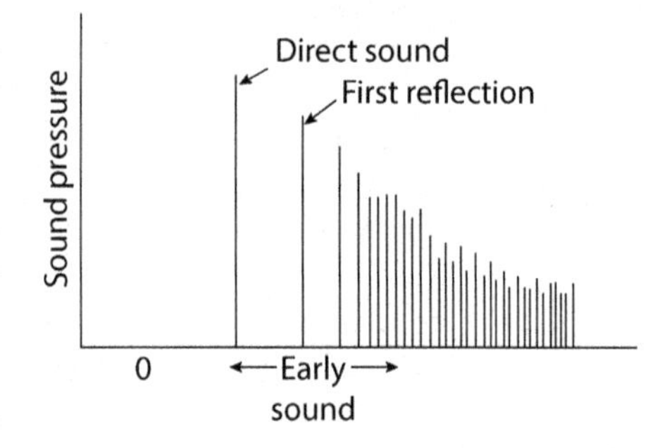

Fig. 16.2. The set of arrived signals is shown.
Adapted from Springer Handbook of Acoustics by Thomas Rossing, 2007 Springer

After the first group of reflections, the reflected sound arrives thick and fast from all directions. The amplitude of these reflections becomes smaller and they follow each other closer in time. This is called **reverberant** sound. If the source produces a continuous signal, the reverberant sound builds up until it reaches some equilibrium level. When the sound stops, the amplitude of vibrations of the air decreases at a more or less constant rate until the sound reaches the limits of inaudibility. In the case of impulse sound, shown in Fig. 16.2, the decay of reverberant sound starts immediately. There is dependence on the time of the decay on the presence of an audience in the concert hall. In the empty concert hall sound decay times are longer than in one that is occupied. This occurs because the absorption of human bodies is pretty big, which is killing the numerous reflections.

The analysis of the acoustics of any auditorium, in a very simple but accurate way, can be obtained from careful study and comparison of direct sound, early reflections, and reverberation.

16.3 The Precedence Effect

The level of direct sound decreases with the distance from the source by 6 dB for each doubling of the distance. Some musical instruments are non-directional: they radiate in essentially all directions; however, some sources are quite directional, such as the brass family.

Our auditory system has an uncanny ability to determine the direction of the sound source, even in the presence of many distracting sounds, such as reflections. For sounds of low frequency, localization depends on the sensation of the very slight difference in time of the arrival of a signal to our two ears. For sounds of high frequency, the difference in sound level at our two ears due to the "shadow" cast by our head provides the main information.

The arrival of the reflected sound from many directions makes the picture more complicated. Our ears usually are provided with several reflections that follow the direct sound from all directions. If these reflections arrive within 50–80 ms after direct sound, our ear does not resolve them as separate signals. These reflections reinforce the direct sound, which is important for listeners sitting at large distances from the source. For sounds that vary rapidly, such as speech, the limit is about 50 ms, for slow music it is closer to 80 ms.

But the most remarkable property of our hearing is that we are able to determine the direction of sound from the very first signal reaching our ear, which is automatically interpreted as the direct path of the sound. This ability of our auditory system is called **precedence effect**. In a nutshell, precedence effect results in the following: the source is perceived to be in the direction from which the first sound arrives.

1. Successive sounds arrive within 35–40 ms.

2. The successive sounds have spectra similar to the first sound.

3. The successive sounds are not much louder (less than 10 dB louder) than the first one.

Let us consider these conditions. If successive sound arrives later than 40 ms after the first impulse, it works like an echo and separate source of sound. If the successive sound does not have essentially the same spectra, for instance, during the symphony concert someone is whispering behind you, it is also disturbing and gives you a clear understanding that you have not one but multiple sources of sound. Also, if the direct sound is too weak, for instance, and you are sitting on the lawn in some semi-open concert place, such as Mann Center in Philadelphia, then the sound from amplifiers standing close can seem direct for you.

All concert halls are built for the full usage of the precedence effect. In small concert halls the early reflections arrive at 20 ms. If the hall has a traditional rectangular shape, early reflections usually arrive from the nearest side wall or from the ceiling. In large concert halls there are always listeners sitting too far from the side walls and ceilings. To provide smoothness in such halls, reflecting surfaces of different types are suspended from the ceiling.

16.4 Properties of Reverberant Sound

The reverberation time is one of the most important acoustical characteristics of any concert hall or auditorium. The behavior of reverberant sound is usually too complicated to be described just by one number, but the reverberation time in the musically important range of frequencies is a good indication of the "richness" of the sound.

Let's consider the steady sound that lasts during time interval T (Fig. 16.3). The signal whose behavior we discuss has the form of rectangle: constant intensity over the time T and abrupt switching off after.

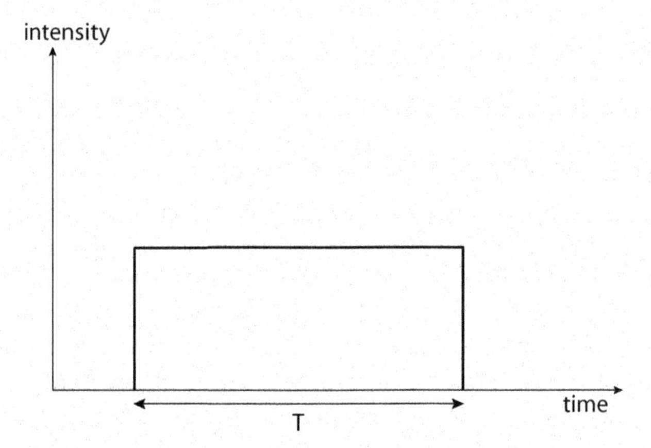

Fig. 16.3. Steady sound which lasts during time interval T.

The signal reaches the listener in a way shown in Fig. 16.4: first comes the direct sound and after it, the series of early reflections. The signal perceived by a listener (Fig. 16.5) does not look like a rectangle anymore. We have some time of gradual growth of the perceived signal, and a gradual decay of sound, after

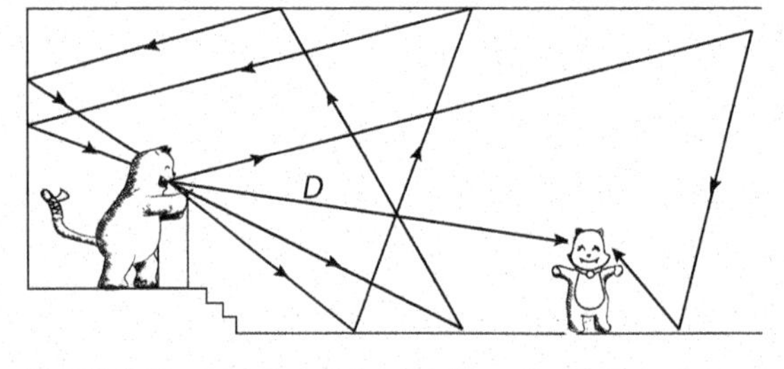

Fig. 16.4. The set of signals which reach the listener.
Adapted from Springer Handbook of Acoustics by Thomas Rossing, 2007 Springer

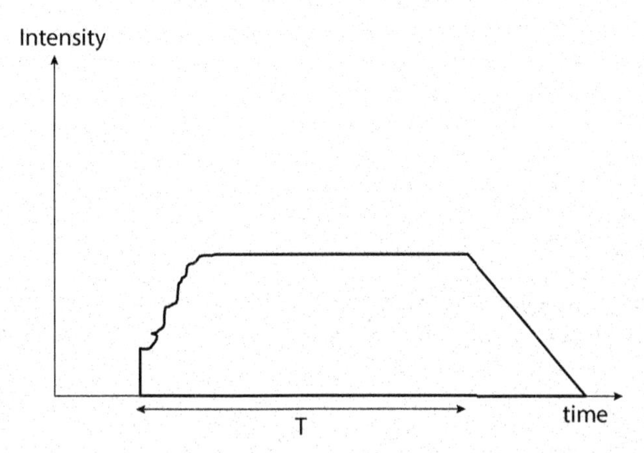

Fig. 16.5. The signal actually perceived by a listener. It does not look like a rectangle anymore.

the signal has been switched off. We can deduce the rates of growths and decay, as well as the reverberant level by considering the sound energy. The source delivers energy, which is living in the airspace of our room, partially absorbed by the walls, the ceiling, and by all objects within the room. The reverberant level is reached when the rate at which energy is delivered by the source is equal to the rate of absorption.

The reverberant sound, together with early absorptions, enhances the direct sound, maintaining the overall loudness. Yet the proper balance between direct, early, and reverberant sound is necessary: the direct sound should be stronger than the background sound—too high a level of reverberant sound may result in a loss of clarity, mixing all the sounds into a total mess.

The dependence of the reverberation time on the size of room can be discussed conceptually with no mathematics involved. When reverberation time is longer, the volume of the room is bihhrt. In a big room, more sound energy can be stored and, as a result, more time is required for the sound to disappear. From the other side, the reverberation time should depend indirectly proportional to the square area of the walls, ceilings, and any other surfaces present in the auditorium. The bigger the square area, the more possibilities exist for sound to be absorbed. Also the reverberation time should depend on the absorbing properties of the materials that cover

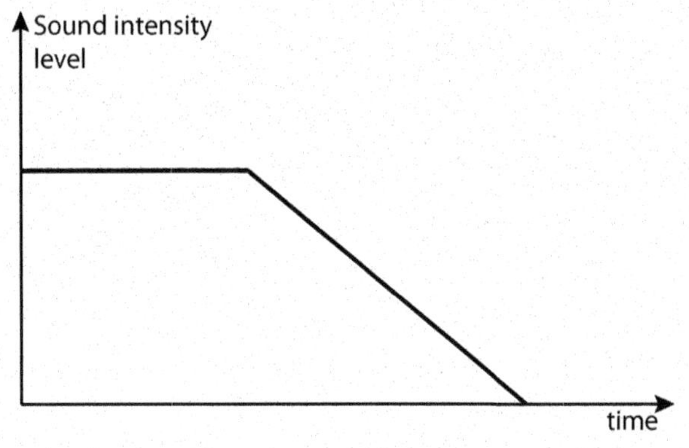

Fig. 16.6. The decay of the sound if sound energy is uniformly distributed over the room.

surfaces: the room with walls covered with felt will demonstrate shorter reverberation times than a marble room.

If the sound energy is uniformly distributed over the room, its decay follows a curve as in Fig. 16.6. Such dependence is called exponential. The reverberation time, exactly like damping time considered in Chapter 10, is defined as the time required for the sound intensity level to decrease by 60 dB.

Sound energy is usually not uniformly distributed over a concert hall or auditorium. As a result, the room can demonstrate not one but two or more characteristic times of decay (Fig. 16.7a)

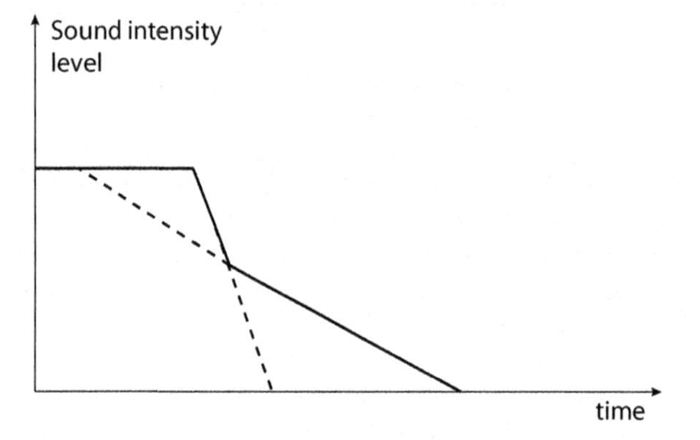

Fig. 16.7a. The room demonstrating two decay times.

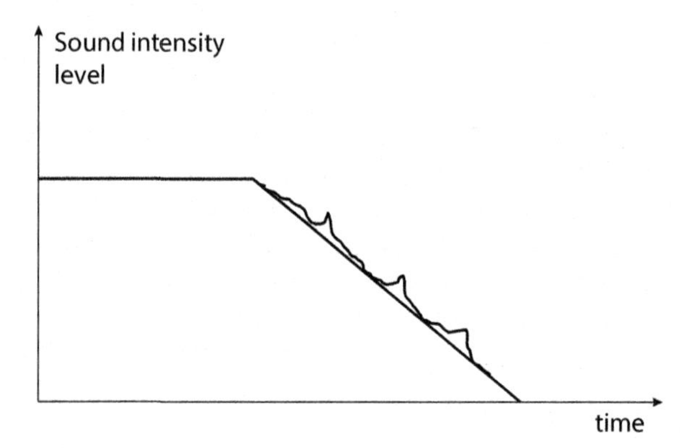

Fig. 16.7b. Decay of signal with some "spikes", showing the existence of additional room resonance.

and the sensation of "dryness" despite the long average reverberation times. The decay curves may also demonstrate some "spikes" (Fig. 16.7b) of an otherwise exponential dependence. These spikes show the existence of the prominent room resonance.

16.5 Calculation of the Reverberation Time: Sabine's Formula

We define the **reverberation time** T_r as the time during which the sound level drops 60 dB below its original level.

Let us discuss the exact relationship between the parameters of the auditorium and the reverberation time. As we already know, the reverberation time should depend on the absorbing properties of the material. It should increase with increasing volume of the room and decrease with increasing the square area of surfaces in the room.

This is summarized in the simple formula first obtained by W. Sabine in 1922 (Sabine's formula):

$$T_r = 0.16 s / m \frac{V}{S_e}$$

where T_r is the reverberation time, V is the volume of the considered room in cubic meters, and S_e is the effective absorption area in square meters. This effective area should be calculated as:

$$S_e = a_1 \cdot S_1 + a_2 \cdot S_2 + a_3 \cdot S_3 + \ldots$$

Table. 16.1. Absorption coefficients of various surfaces for musically important frequency ranges.

Material	frequency (Hz)					
	125	250	500	1000	2000	4000
Acoustical Tile, suspended	0.75	0.93	0.83	0.99	0.99	0.94
Acoustical Tile, on concrete	0.14	0.2	0.76	0.79	0.58	0.37
brick	0.03	0.03	0.03	0.04	0.05	0.07
carpet, on concrete	0.02	0.06	0.14	0.37	0.6	0.65
carpet, on pad	0.08	0.24	0.57	0.69	0.71	0.73
Concrete Block, unpainted	0.36	0.44	0.31	0.29	0.39	0.25
concrete Block, painted	0.1	0.05	0.06	0.07	0.09	0.08
Draperies, mediumweight	0.07	0.3	0.5	0.7	0.7	0.6
glass, window	0.3	0.2	0.2	0.1	0.07	0.04
heavy plate glass	0.2	0.06	0.04	0.03	0.02	0.02
platform floor, wooden	0.4	0.3	0.2	0.2	0.15	0.1
uphostered seeting, unoccupied	0.2	0.4	0.6	0.7	0.6	0.6
upholstered seating, occupied	0.4	0.6	0.8	0.9	0.9	0.9
wood/metal seating, unoccupied	0.02	0.03	0.03	0.06	0.06	0.05

where α_n are the coefficients of absorption of the particular material and S_n is the square area of this particular surface (for example, S_1 for the floor, S_2 for the walls, S_3 for the ceiling, etc.).

The absorption coefficients of various surfaces are shown in Table 16.1. These coefficients are in the range from 0 to 1 and represent the fraction of the incidental sound energy that is lost for a listener in each reflection. The more rigid a material is (see brick, for example), the less the absorption coefficient is.

A material is called a perfect reflector if its absorption coefficient is equal to zero. In such a hypothetical situation S_e in the Sabine's formula is also zero. No energy is taken from the reverberant sound by the absorption. Zero in the denominator of any fraction means that the result of the division is equal to infinity. So, if we managed to create the room with all surfaces of perfect reflectors, the sound in such a room would never die away. The material with absorption coefficient 1 is called a perfect absorber; it reflects nothing from the wall, and sound decays in such a room momentarily. A good example of a perfect absorber is an open window: all the sound runs away, and nothing comes back.

Note that what stands in the denominator of Sabine's formula is a product of effective surface area and absorption coefficient. An area of 10 m^2 covered by material with an absorption coefficient of 0.2 has the same acoustical properties as 5 m^2 element of surface with absorption coefficient 0.4.

Example 16.1. The rectangular room has the volume of 100 m^3 and surface area of 200 m^2. It is all covered with a material with coefficient of absorption of 0.1. Calculate the reverberation time.

Answer:

Following Sabine's formula,

$$T_r = \left(0.16\frac{s}{m}\right)\frac{V}{S_e} = \left(0.16\frac{s}{m}\right)\frac{100m^3}{0.1 \cdot 200m^2} = 0.8\,s.$$

Example 16.2. The rectangular room has the volume of 100 m^3 and surface area of 200 m^2. Half of this area is covered with material of absorption coefficient 0.1 and the other half with material of absorption coefficient 0.2. Calculate the reverberation time of this room.

Answer:

Following Sabine's formula,

$$T_r = \left(0.16\frac{s}{m}\right)\frac{V}{S_e} = \left(0.16\frac{s}{m}\right)\frac{100m^3}{\left(0.1 \cdot 100m^2 + 0.2 \cdot 100m^2\right)} = 0.53\,s.$$

Example 16.3. During remodeling the material with coefficient of absorption 0.1 covering some of the room was substituted with the material with coefficient of absorption 0.2. How did it influence the reverberation time?

Answer:

The increase of denominator in the Sabine's formula twice leads to decrease of reverberation time, also twice.

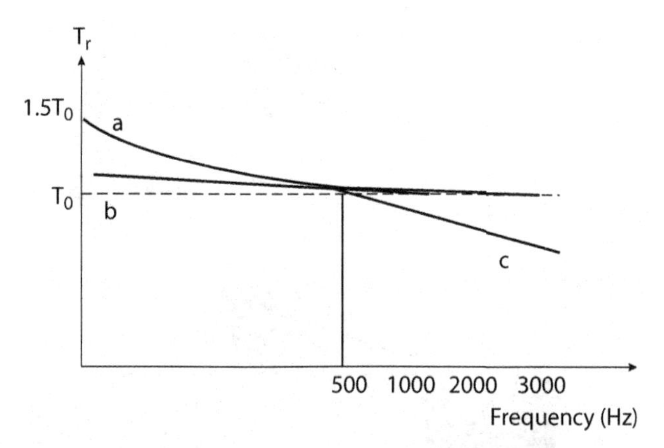

Fig. 16.8. Reverberation times as a function of frequency. The most desirable dependence corresponds to (a). Curve (b) shows too short times in low-frequency range, meanwhile, (c) in high frequency range.

16.6 Dependence of Reverberation Time on Frequency

When we talk about a single number for the reverberation time of some particular room, we usually refer to the mid- and high-frequency ranges, 500 Hz and above. The reverberation times usually gradually increase toward the low-frequency end so that they are approximately 50% longer in the extreme bass than in the treble, as can be seen in Fig. 16.9. The most desirable dependence of the reverberation time (curve a) gives approximately constant reverberation time for the frequencies above 500 Hz and substantially longer times for the low frequencies. Fortunately many building materials have lower absorption coefficients at low frequencies (see Table 16.1).

Note that a 60 dB level in the bass range represents a greater decrease in loudness level (phons, see Fletcher-Munson diagram in Chapter 7) than in treble, so equal reverberation times at all frequencies would create for your ears a sensation that the bass has dropped too fast. Too-short reverberation times at low frequencies causes the room to lack liveliness of the sound (Fig. 16.8, curve b). If the high-frequency reverberation is too short, the brilliance is lacking (Fig. 16.8, curve c).

16.7 Reverberation Times for Rooms of Different Purposes

The desired length of the reverberation time strongly depends on the purposes of the particular room. As can be seen from Fig. 16.9, the reverberation time increases with the increase of the size of room, which was expected from Sabine's formula.

Speech and organ music are extremes: speech requires the shortest possible reverberation time, meanwhile, the richness of the organ music relies on the longest reverberation times. This makes it particularly difficult to create the appropriate acoustics in the cathedrals.

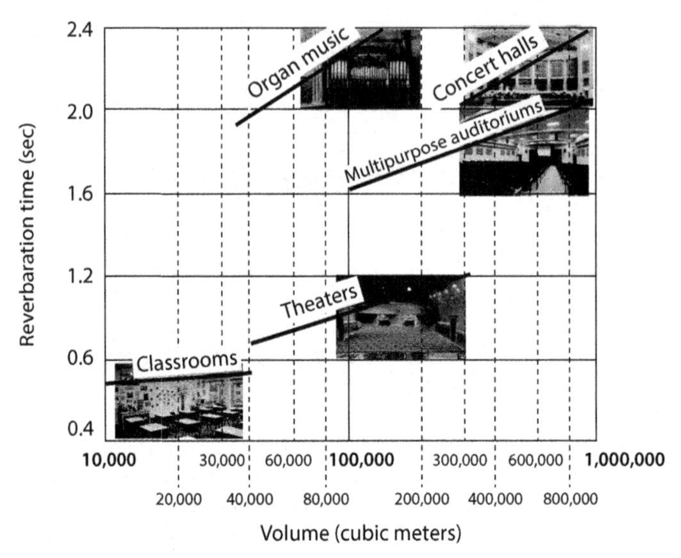

Fig. 16.9. Reverberation times required for rooms of different purposes. Note dependence of reverberation times on the volume and surface of rooms.

The numbers shown in Fig. 16.9 are only typical; the individual preferences of the performers can be 10–20% higher or lower. Even the same room can require different reverberation time: baroque music, for instance, needs shorter times than romantic music does.

The longest reverberation times are required for organ halls, which all are all different and require some sophistication from the performers switching from one hall to another. American organ halls usually have a reverberation time of 1–2 s; the legendary Mormon Tabernacle in Salt Lake City has one of the longest times of 3.5 s in its acoustical properties. Meanwhile, for European organ halls, reverberation times of 6–8 s are pretty typical.

As can be seen from Fig. 16.9, a multipurpose auditorium demonstrates the reverberation time suitable for music but too long for speech. Usually in such halls the necessary compromise has been reached by the electronic sound reinforcement.

Summary, Terms, and Relations

General requirements for the room serving for musical purposes: uniformity, clarity, envelopment, freedom from echo and noise, proper reverberation time.

Precedence effect: our auditory system defines the location of the source based on the very first arrival of the sound, even if this signal is up to 10 dB smaller than the consequent one.

The reverberation time is the time during which the sound level drops 60 dB below the original level. The reverberation time should be calculated following Sabine's formula:

$$T_r = 0.16\,s\,/\,m\frac{V}{S_e}.$$

Questions and Exercises

1. Discuss the acoustics of your auditorium. What details that serve the acoustical purposes can you see immediately in the interior? (Hint: Look at the ceiling as well as curved walls near the blackboard. Check if your auditorium has a slope.)

2. Estimate the reverberation time of your auditorium.

3. If you stand at one side of a stage with hard parallel side walls 17 m apart and clap your hands once, how often do the echoes arrive?

4. What is the biggest difference between reverberation time for American and European organ halls?

5. What would happen to the reverberation time if all covering materials were substituted with materials of double the absorption coefficient?

6. Consider a hollow cube with sides of 5 m. If all sides of this cube have the absorption coefficient 0.2, what is the reverberation time?

7. If a 20 x 17 x 5 m room has a reverberation time of 1 s, what is the effective surface area? Estimate what the coefficient of absorption should be.

8. Estimate the reverberation time for frequency 1000 Hz of a typical living room 5 x4 m with a ceiling 3 m high. The floor is hardwood, and the ceiling and walls are plaster.

APPENDIX A

Metric System

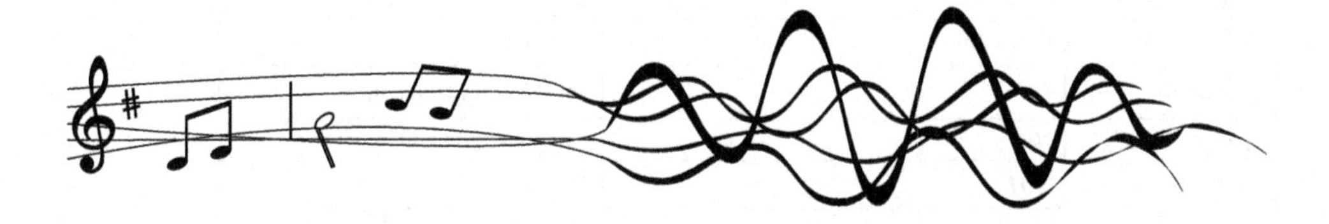

A.1 The Basic Metric Units

In Physics we do not use the units of the Imperial System (lbs, ft, inches). It is interesting that the agreement about units of length and mass took such a long time. At the end of the eighteenth century, practically every country had its own units of mass and length. There were plenty of feet (all different) and plenty of pounds (also different). And almost every country considered its units as the most important ones. The French Revolution brought us not only the ideas of liberty and equality but also the basic units of the now-well-known International System: meter and kilogram.

These units were the first units that weren't connected to the measurements of particular royal people (foot, for instance), but had strong connections to our common home—planet Earth. As the unit of length, one ten-millionth of the distance between the North Pole and the equator was established. The unit of mass, 1 kilogram, is just a mass of 1 liter of the most common liquid on our planet—fresh water.

Table A.1 shows the correspondence of several important physical concepts in the Imperial and Metric systems.

Table A.1. Correspondence of physics units in Imperial and Metric system.

Concept	Metric Unit	Imperial Equivalents
length	1 m (meter)	3.18 ft
mass	1 kg (kilogram)	2.2 lbs
velocity	1 m/s	2.24 mph
force	1 N (newton)= 1 kg-m/s^2	0.225 lb
pressure	1 Pa (pascal) = 1 N/m^2	1.46 $\times 10^{-4}$ lb/in^2
energy	1 J (joule) = 1 N-m	0.74 ft-lb
power	1 W (watt) = 1 J/s	1.35 $\times 10^{-3}$ hp

A.2 Prefixes in the Metric System

Another advantage of the metric system is the use of prefixes. It is not convenient to measure the speed of a snail, say in m/s, it will be, rather, in mm/s (millimeters per second). Prefixes are the same for all concepts: meters, seconds, grams. In Table A.2 the most important prefixes are shown with symbols and examples.

Table A.2 Standard Prefixes

Prefix	*Equivalent Multiplier*	*Example*
μ = micro	10^{-6}	1 μm = 0.000001 m
m = milli	10^{-3}	1 mg = 0.001 g
c = centi	0.01	1 cm = 0.01 m
K = kilo	10^3 = 1000	1 kg = 1000 g
M = mega	10^6	1 MHz = 10^6 Hz
G = giga	10^9	1 GHz = 10^9 Hz

A.3 Power of Ten Notation

The numbers that appear in measurements are pretty often either too big or too small, so we should use a lot of zeros either after the significant figures or before them, after the decimal point. As a result, scientists often use a shorthand notation that is based on multiplying or dividing by 10. For example, the number 10000 = 10x10x10x10, so we could write it as 10^4. The number 0.00001 = 1/100000 = 10^{-6}.

Other numbers can be written in such form, too. For instance, 53,500,000 = 53.5 x 10^6 or 6.35x10^7. This also applies with negative exponents; for example, 0.00034= 3.4/10000 = 3.4x10^{-4}.

Numbers that are written in the notation of the power of ten can be divided or multiplied very easily. For example, we need to multiply 2x10^6 and 3x10^6.

$(2\text{x}10^6)\text{x}(3\text{x}10^5) = 2\text{x}3\text{x}10^{(6+5)} = 6\text{x}10^{11}$.

You do not need a calculator for this!

Let's now divide two numbers; for example, 6x10^7 by 3x10^2:

$(6\text{x}10^7)/(3\text{x}10^2) = (6/3)\text{x}10^{(7-2)} = 2\text{x}10^6$.

To conclude,

Multiplication can be written in a general form as:

$(Xx10^n)x(Yx10^m) = (XxY)x10^{n+m}$;

Division can be generalized as:

$(Xx10^n)/(Yx10^m) = (X/Y)x10^{n-m}$.

Appendix B

Written Music

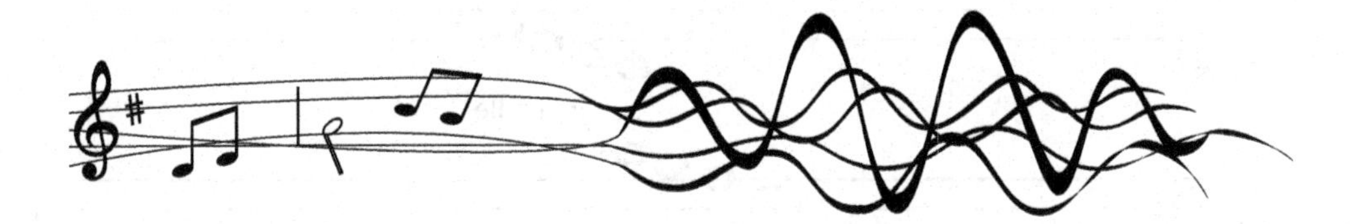

In standard Western musical notation, the staff, or stave, is a set of five horizontal lines and four spaces that each represent a different musical pitch. The history of these five lines is not very simple; there is a legend that initially 11 lines were used (Fig. B.1). Each line and each space between is a place for the letter-named note of the white keys of the piano. I don't know how and why A became A, but the notes follow from A to G, after G–A again, emphasizing the fact that two notes one octave apart are identical in character and color. This group of eleven lines has a note C_4 in the middle, so C_4 is often called the "middle C." The line for middle C disappeared, splitting the grand staff into two sets of 5 lines each (Fig. B.1). The remains of this disappeared eleventh line, which serves for C_4, are called the ledger line.

It is a little bit confusing, but octaves start from Cs, not from As. So, the note one octave lower than C_4 is C_3, starting the octave D_3, E_3, G_3, A_3, B_3 (Fig. B.2).

Each group of five lines uses the bass clef (Fig. B.3a) and treble clef (Fig. B.3b) because, as we can see from Fig. B.4, the notes can have the same position on the lines but different meanings (compare, for example, C_3 of bass clef and A_4 of treble clef).

The black keys of a piano are addressed with special symbols: a sharp sign #, which means "the key adjacent to the corresponding white key to the right," or one semitone higher (Fig. B.2); and a flat sign ♭, which means "the key adjacent to the given key to the left," or one semitone lower (Fig. B.2).

The keyboard of a piano with frequencies of notes is shown in Fig. B.6.

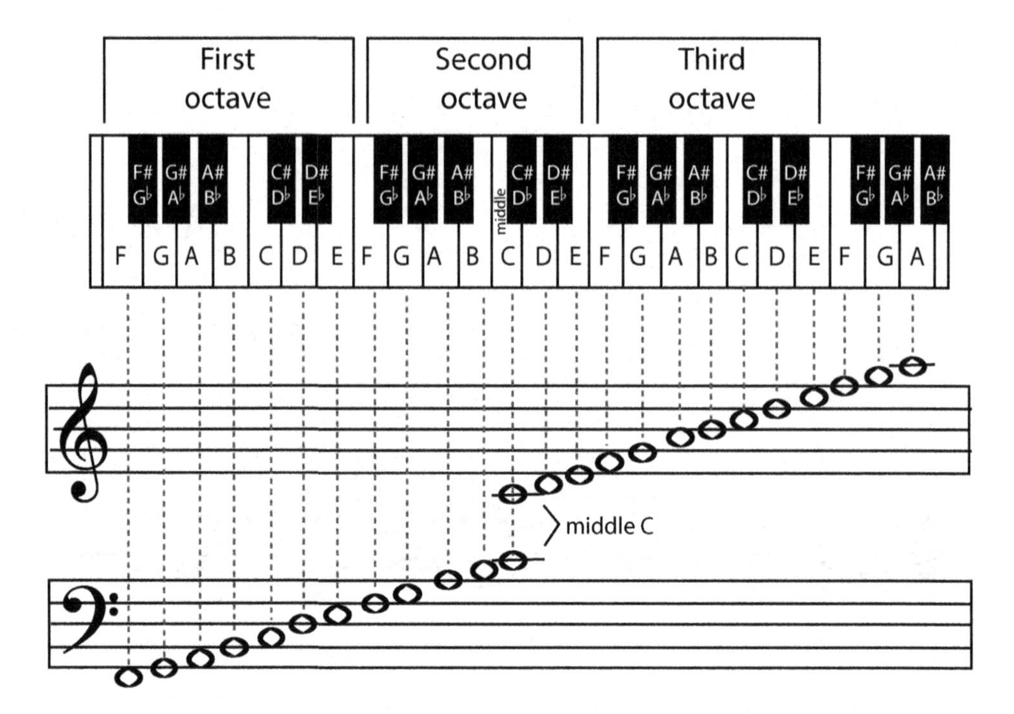

Fig. B.1. Grand staff: positions of white keys are on and between lines. The middle "C" is placed on the ledger line. The actual keys on the piano are also shown.

Fig. B.3a. Bass clef.

Fig. B.3b. Treble clef.

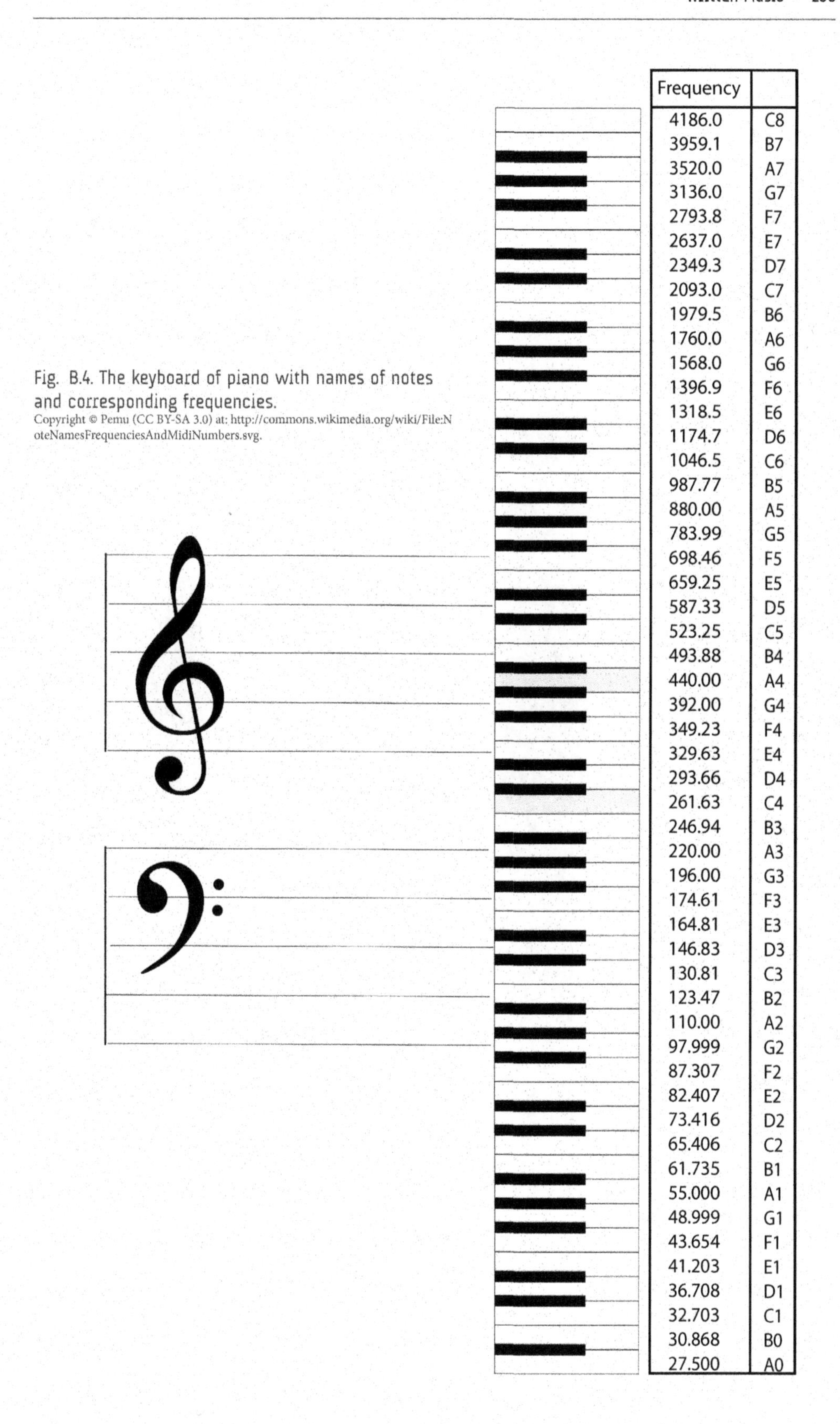

Frequency	
4186.0	C8
3959.1	B7
3520.0	A7
3136.0	G7
2793.8	F7
2637.0	E7
2349.3	D7
2093.0	C7
1979.5	B6
1760.0	A6
1568.0	G6
1396.9	F6
1318.5	E6
1174.7	D6
1046.5	C6
987.77	B5
880.00	A5
783.99	G5
698.46	F5
659.25	E5
587.33	D5
523.25	C5
493.88	B4
440.00	A4
392.00	G4
349.23	F4
329.63	E4
293.66	D4
261.63	C4
246.94	B3
220.00	A3
196.00	G3
174.61	F3
164.81	E3
146.83	D3
130.81	C3
123.47	B2
110.00	A2
97.999	G2
87.307	F2
82.407	E2
73.416	D2
65.406	C2
61.735	B1
55.000	A1
48.999	G1
43.654	F1
41.203	E1
36.708	D1
32.703	C1
30.868	B0
27.500	A0

Fig. B.4. The keyboard of piano with names of notes and corresponding frequencies.

References

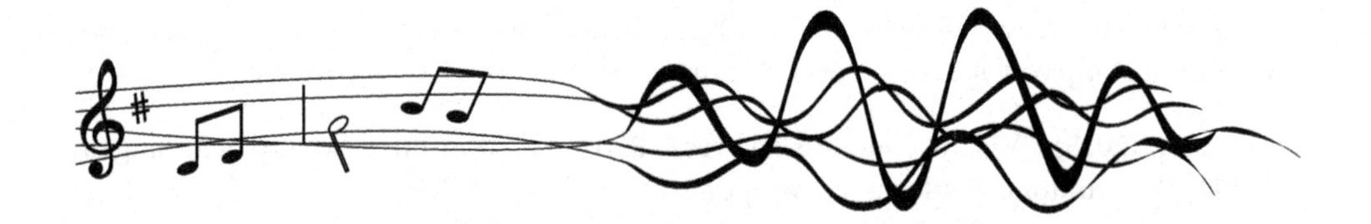

Anderson, G. (2010). "208th Army Reserve Band spread holiday cheer." U.S. Army. Retrieved from http://www.army.mil/article/49694/208th_Army_Reserve_Band_spread_holiday_cheer/

Cassidy, D., Holton, G., & Rutherford, J. (2002). *Understanding Physics*. Springer-Verlag: Heidelberg, Germany.

Denes, P. & Pinson, E. (1963). *The Speech Chain*. Bell Telephone Laboratories.

Dobre, A. & Ramtal, D. (2014). Physics for JavaScript Games Animation and Simulations with HTML5 Canvas. Apress: New York, New York.

Foley, H. J. & Matlin, M.W. (n.d.) *Sensation and Perception*. Retrieved from http://www.skidmore.edu/~hfoley/Perc11.htm

Forinash, K. (n.d.) "Sound: An Interactive eBook on the Physics of Sound." Indiana University Southeast: New Albany, IN. Retrieved from soundphysics.ius.edu

Gough, C. (2014). "Musical Acoustics", Springer Handbook of Acoustics. Springer-Verlag: Heidelberg, Germany.

Hall, D. (2002). *Musical Acoustics*, 3rd ed. Brooks/Cole.

Hassan, O. A. B. (2009). "Room Acoustics", Building Acoustics and Vibration Theory and Practice. World Scientific: Singapore.

Holoman, D.K. (1998). *Masterworks*. Prentice Hall. Retrieved from http://cwx.prenhall.com/bookbind/pubbooks/masterworks/medialib/fundamentals/voices02.html

McCall, M. W. (2010). "Oscillatory Motion." *Classical Mechanics. From Newton to Einstein: A Modern Introduction.* Wiley & Sons, Ltd: West Sussex, United Kingdom.

Meyer, J. (2009). *Acoustics and the Performance of Music: Manual for Acousticians, Audio Engineers, Musicians, Architects and Musical Instrument Makers.* Springer-Verlag: Heidelberg, Germany.

Nave, R. (n.d.). "HyperPhysics." Georgia State University. Retrieved from hyperphysics.phy-astr.gsu.edu

Paranjape, A. A., Chung, S. & Selig, M.S. (2011). "Flight mechanics of a tailless articulated wing aircraft." *Bioinspiration & Biometrics, 6*, University of Illinois: Urbana, Il.

PlaneTuning (2015). "Violin 'Loudness Curves.'" Retrieved from http://www.platetuning.org/CMH_vln_semitone_loudness_curves.pdf

_______. "The Physics Forum." Retrieved from www.physicsforums.com

Rossing, T. D., (2014). *Springer Handbook of Acoustics*, 2nd ed. Springer-Verlag: Heidelberg, Germany.

Syncopation. (n.d.), In *Wikipedia*. Retrieved from http://en.wikipedia.org/wiki/Syncopation

Veldman, A.E.P. (n.d.) "Computational Fluid Dynamics at RuG." University of Gronigen. Retrieved from http://www.math.rug.nl/~veldman/cfd-gallery.html

Vibrationdata (n.d.). Piano Page. Retrieved from http://www.vibrationdata.com/piano.htm

von Bekesy, G. (1960). *Experiments in Hearing.* McGraw-Hill.

Wright, T. & Gerhart, P. (2009). "Turbomachinery Noise", F*luid Machinery Application Selection and Design, 2nd ed.* CRC Press: Boca Raton, FL.